Hard to Swallow

Hard to Swallow

The Truth About Food Additives

Doris Sarjeant

Karen Evans

Published by *alive* **books**
7436 Fraser Park Drive
Burnaby BC Canada V5J 5B9
(604) 435-1919 or (800) 663–6580

Editing: Paul Razzell
Cover Photo: Edmond Fong
Typesetting/Layout: Scott Yavis, Raymond Cheung
Cover Design: June Vance

First edition – February 1999
Second edition – August 1999

Canadian Cataloguing in Publication Data
Sarjeant, Doris D. (Doris Doreen), 1947–
Hard to swallow

Previous ed. has title: The truth about food additives.
Includes bibliographical references.
ISBN 0–920470–47–5

1. Food additives. 2. Food additives – Law and legislation – Canada. 3. Radiation preservation of food. 4. Food – Biotechnology. 5. Food – Labelling – Law and legislation – Canada. I. Evans, Karen S. (Karen Suzanne), 1953- II. Title III. Title: The truth about food additives.
tx553.a3s27 1998 664'.06 c98–910766–3

Printed and bound in Canada

1 3 5 7 9 10 8 6 4 2

To God, who promises:

he who has begun
a good work in you, will perform it until the end

- Philippians 1:6

Contents

Foreword

by Ross Hume Hall, PhD
Professor Emeritus, McMaster University

The safety of a food additive is not something you can personally check. It takes sophisticated laboratories and technical knowledge to do the testing. So you depend on the government to ensure that the testing is done. Yet, the federal government allows thousands of chemical additives in your food that have not been tested for safety. Moreover, the government enforces a set of policies that make it nigh impossible to remove those untested chemicals. The equation of people's rights versus chemical rights tilts scandalously in favour of chemicals.

How did I arrive at this conclusion? It certainly did not happen overnight. In fact, it was the result of a personal odyssey that lasted several years. It began innocently with a letter. I noticed among the ingredients listed on a product label a chemical additive that I knew from the medical literature was darkened by a cloud of suspicion. What was the chemical doing in this product? I felt that at least it shouldn't be allowed in any food until proven safe and all suspicions dissipated. I wrote the head of the Health Protection Branch (HPB) at Health Canada, the Ottawa agency responsible for overseeing the safety of food additives. I innocently expected if I reminded him of the black cloud hanging over the food additive he would set the wheels in motion to recall this additive. Or perhaps there was no need to set wheels turning, HPB had some information I didn't know that showed the additive to be safe.

The reply came, an obvious form letter. I had a vision of some bottom-level clerk pulling Form Reply No. 12 out of the file and typing my name onto the head. The additive is safe, the letter said, and if I had any proof to the contrary they would consider it. The file clerk didn't respond to the fact the medical community regarded the chemical with suspicion, nor did the clerk offer any reason why HPB believed it to be safe.

I thought HPB's attitude weird, particularly the part about asking me to prove the additive unsafe. They have scientists who can read the medical articles detailing the potential dangers of this additive. They have the facilities and the resources to evaluate its safety, yet an unsavoury attitude wafted from that letter, as appealing as a sun-ripened fish. They were putting the onus on me to come up with the evidence to prove this particular additive unsafe.

I learned later that, if you try to prove an additive to be unsafe, you have as much hope of success as climbing Mount Everest in your underwear. You'll be drawn into a hearing. You'll have to produce shelf-loads of written documents. You'll have to hire lawyers to put your argument into legal language. You'll have to quit your job and devote full time to your case. You'll be cross-examined by lawyers for the company that makes the additive, who will prove that you are an unworthy scoundrel not to be trusted, wasting everyone's time.

The government's policy on the safety of chemical food additives is: no evidence of harm, we approve – and it stays approved. Here's why. In the 1920s and 1930s, when the majority of additives permitted in Canada's food were first approved, the science of testing chemicals for safety was crude and incapable of detecting most dangers. Thus, failure to detect danger did not necessarily mean the additive was benign.

The situation reminds me of the time before they installed metal detectors in airports. Terrorists could carry revolvers, grenades and bombs on board an aircraft and never be caught. In a similar way the laboratory testing methods of 1930 didn't stop dangerous chemicals from finding their way into foods.

Since 1930, the sophistication of safety testing has shot skyward. We have a fantastically better idea now of the ease with which foreign chemicals, such as food additives, can disrupt the body's workings. We could design and set up chemical safety-testing programs equivalent to putting metal detectors in airports. With these new tests we could reevaluate the current list of food additives and screen out the worst offenders.

I say could, but it doesn't happen. The government has no intention as far as I can see of using this modern knowledge. The knowledge lies dormant on library shelves. Hence, when you take a bite of commercial food containing food additives, foreign chemicals enter your body whose safety was approved with 1930s knowledge.

When I realized that HPB was not applying this new knowledge, I couldn't believe it. Why doesn't the government use this new information for the benefit of consumers? Doesn't the public deserve protection? I set out to find why the government's methods of testing the safety of food additives are stuck in the 1930s. This was my odyssey.

* * *

Before getting into the bowels of Ottawa, however, we ought to look at some of the new scientific knowledge about ways foreign chemicals can harm your health, particularly if you are a woman or a child.

Scientists finally discovered women and men are different. Don't laugh. Back in the 1920s and 1930s the standard belief was that a human is a human. They tested candidate food additives with the male body in mind – young, robust, healthy males at that. Government testers estimate the amount of a foreign chemical healthy young men can tolerate then set the safety level about ten times lower, assuming this "safety factor" protects women. We now know this assumption to be outrageously wrong. A woman's body with its sensitive reproductive system and unique hormonal cascades is much more vulnerable to disruption by foreign chemicals than is a man's body – as much as 1,000 times more vulnerable. The vulnerabilities range from a lowering of immune defenses to a greater susceptibility to cancer.

As for children, government approvers of food additives act as if all Canadians are twenty-one years old at birth. Fetuses, toddlers and children simply don't exist in government thinking. Imagine how defenseless a baby or child is against the blows of a deranged molester. We know now that children are just as defenseless against chemicals. The threshold of vulnerability to chemical assault is far lower in children than in adults and the threshold is lowest of all in the unborn.

The placenta, unfortunately, offers no shield. A pregnant woman, for instance, eats a burger with food additives. These foreign chemicals enter her blood and, like little knives, zip through the placenta straight into the naked fetus.

Of all the stages of life, the first nine months after conception unfold in the most wonderfully intricate manner, as a petite cluster of cells flowers into a human form. But the very intricacy of the rapid changes makes it easy to block or wrongly channel the flowering. You are familiar with the fact certain chemicals cause birth defects. When we talk of defects, we tend to think of visible impairments, such as missing limbs or defective facial features, obvious and unfortunate. Just as insidious and just as devastating are the hidden impairments. Our understanding of how foreign chemicals cause hidden birth defects has advanced enormously in the last two decades. Many defects are not what you would call life threatening but nevertheless can disable an individual for life.

Take the human brain, for instance. It forms mainly during the fetal stage. Two things are happening: neurons are being laid down, a thousand a second; and the neurons are being wired to each other, forming a complex network that becomes the individual's nervous center and thinking being. This hard wiring – called hard because once in place it's there for life – is under delicate hormonal control, easily altered or disrupted by foreign chemicals.

Disruption of the development of the human brain during the fetal stage can lower intelligence and cause a measurable drop in short term memory later in life. The individual still functions but, sadly, is denied the potential inscribed in his or her genes at the conceptual union.

That's not all. Chemical exposure during the fetal stage can set in motion tissue changes that years later, as an adult, erupt into cancer.

I don't wish to leave the impression that all foreign chemicals to which the fetus is exposed cause these nasty problems. Some do, some don't, but until tests are carried out we won't know which are which. None of the chemicals currently approved as food additives have ever been tested for their full potential to harm the unborn.

* * *

Up to the time I wrote that letter to HPB, the government's antiquated approach to safety testing food additives escaped me. I didn't know how the government went about approving food additives and I didn't care. I ate a reasonably balanced diet with a fair share of commercial food products loaded with additives. Blithely, I assumed the government wouldn't allow supermarkets to sell anything deemed unsafe.

I should have known better. I had detailed knowledge how the human body works. Moreover, I was researching how foreign chemicals – in this case drugs – damage human cells. I talked to colleagues around the world doing similar work. I had my fingers on the latest knowledge about body frailties. But like a lot of scientists, I divided my life into compartments. I didn't see the relevance of my professional life to my home life. It never occurred to me that my research had a bearing on the food I ate at home.

The letter episode changed all that. Foods I normally ate with gusto suddenly took on a chemical hue – at least to my chemist's eyes. What would these chemical additives do to my body or the bodies of the members of my family? I began to wonder just how the government goes about approving food additives, pesticides and other chemicals allowed in the Canadian foods. Most troubling, why weren't they using the latest scientific knowledge I knew was pouring forth from the world's laboratories? But I was a babe in the lion's den. I had no idea how the Ottawa bureaucrats make decisions, nor did I have any idea of how to find out.

The HPB of Health Canada is located in an area of Ottawa called Tunney's Pasture. I imagine a farmer named Tunney once grew crops there. The area doesn't look like a pasture anymore. Dozens of red brick buildings have sprouted, each patrolled by uniformed guards. You can't go to Tunney's pasture and knock on a door and say: "Pardon me, but could you tell me how you make decisions about foreign chemicals allowed in my food." For a private citizen, Tunney's pasture and the food bureaucracy are as impenetrable as a thicket of thorns.

About that time, quite independent of an aroused interest in food additives, a lucky break came. I was appointed to the Canadian Environmental Advisory Council (CEAC). You've probably never heard of this group. Mouse-quiet, this council operates behind the scenes, never giving out public statements. The council's ten or so citizens, drawn from across the country, have a sole mission: advise the federal Minister of the Environment. I was appointed because of my knowledge of how foreign chemicals can affect the human body.

The appointment for me was like going back to university. I was about to get an education in Chemical Politics 101. The thicket opened. Suddenly I was catapulted into the highest decision-making levels of the federal government. Our council members talked frequently to the minister, the deputy minister and other senior officials. We were privy to behind-the-scenes decisions, the sort of things that triggered a raised eyebrow of the minister. My tales of how foreign chemicals destabilize human health seldom, if ever, raised an eyebrow. The raised eyebrow concerns were: first, about the minister's public image and, second, how special interest groups, like big corporations, would react to proposed policies. Other concerns , like human health, trailed off to the bottom of the agenda.

That was only part of my education. Learning about chemical politics went beyond meetings with the minister. We needed information to give advice and the best way of getting that information was to interview bureaucrats. We could talk to bureaucrats at all levels in the environment department as well as other government departments. This privilege gave me the ability to float, so to speak, through the entire government bureaucracy, including the hitherto impenetrable HPB in Tunney's Pasture. No door was closed.

Early on I interviewed one of the bureaucrats in his office. I wanted to know what happens when new scientific data comes along that proves a food additive, for instance, is harmful to the fetus. Would the new information automatically cause the government to withdraw the food additive?

"Not necessarily," he said. "It is up to the minister."

"What do you mean up to the minister," I said. "She's a politician, what does she know about the technical details of why this food additive is harmful? If the foreign chemical is a proven danger to the fetuses of the country why doesn't she just say 'ok, cancel approval'?"

"That's precisely the point," the bureaucrat said looking at me as if I were the dimmest person ever to enter his office. "She's a politician. The final decision to approve or not to approve has to be hers. Oh sure the technical staff advises her but she makes the decision. She has to weigh the politics of the issue."

"What politics? This is a scientific issue, a matter of harm to babies. The facts are clear. The food additive is a menace."

"The facts may be clear to you, but there are other scientists who claim the food additive is not so dangerous."

"Sounds to me these are the same scientists that claim its unclear that tobacco smoke is harmful."

Ignoring my remark, the bureaucrat said before the minister makes a decision, she weighs the consequences of withdrawing the food additive from the market. "Is there a good substitute? She'll hear from the food companies. Maybe workers will lose their jobs. She has to weigh all these matters."

"And the babies that will continue to be exposed to this dangerous chemical?"

Rattling his pencil box, the man said, "Babies are just part of the equation that goes into the decision."

He picked out a red pencil and started writing a note in the margin of a paper on his desk.

* * *

I know it's wishful thinking, but I often wish that government decisions about permitting foreign chemicals in our food were not based on politics but based on the way the human body works.

In my university classes, I taught biochemistry and nutrition. I'd tell my students, "look at food from your body's point of view. You eat food every day and your body has to deal with every gram of it. Your body can't set the food aside and say I'll deal with it tomorrow; it's got to deal with it now – and fast. Your body recognizes only two categories of substances: substances that nourish and substances that poison.

Substances that nourish, as soon as they enter the digestive tract, give you the energy you need, renew worn-out tissues and, of course if you are a child, promote growth.

Your body treats substances that can't be used for nourishment as foreign, a potential to poison. To poison means to interfere with the way your body changes food into energy or renewed tissue, a biochemical process called metabolism. Metabolism is a beautifully choreographed set of actions as intricate as a ballet. Substances that interfere are like marbles. Throw them under the dancers' feet, arms flail and coordination falters. The ballet becomes a mess. In the same way, the instant you eat a food containing substances the body can't use for nourishment, those substances are poised like marbles for throwing on the stage floor.

Notice I said poised, because once the body recognizes it can't use the substance for nourishment it has to do something fast, before the marbles, so to speak, are thrown. The body has a number of clever ways to do this. The body can change the substance so the kidney can discharge it into the urine. Sweat also carries off a lot of noxious substances. (The health benefits of the sauna have long been recognized.) The body even deposits some foreign chemicals in hair. As the hair grows it'll eventually be cut off along with any nasty residue.

If all these routes fail and the body is still stuck with this foreign substance, there's one last resort, lock up the substance in body fat. It's not the best route because the poison is still in the body. At least it is out of the way, like storing things in your attic. All in all, your body puts up a staunch defense when dealing with foreign substances from your food.

Sometimes that defense takes an ironic twist. One of my colleagues, a woman, had five children. She breast-fed all the kids. During her own childhood she had lived near a chemical dump and had eaten food contaminated with polychlorinated biphenyls (PCBs). Polychlorinated biphenyls have been associated with a number of serious health problems, including mental retardation and cancer, and are among those foreign chemicals that lodge in body fat. My colleague, a biologist and a practicing midwife, knew that year by year her body fat was becoming loaded with polychlorinated biphenyls. When she fed her first baby, she estimated about one third of her PCB body burden passed into the milk and into the baby. By the time she nursed five kids her body had been completely detoxified.

She knew exactly what was happening: her body was transferring the PCBs from her fat to the fat in her milk. She benefited, but her babies? On balance, she felt the advantage of breast-feeding outweighed the disadvantage of baby ingesting polychlorinated biphenyls. Her children are now grown, yet she worries about the potential long-term effects of these chemicals. Was it the right choice? It's not a choice a mother should have to make.

Regardless of how your body deals with foreign chemicals, there's a price to pay. Foreign substances, like food additives, erode your body's defenses, allowing at first a few marbles to get thrown, then more and more as the years go by. Your metabolic ballet looks more and more shabby and uncoordinated.

What does this mean to your long-term health? First of all, your immune system weakens. You become more vulnerable to minor ailments, colds and sore throats. Body tissues don't function as well. Cancer and cardiovascular problems can arise. When your metabolism isn't efficient your body is vulnerable to practically every ailment known in medical textbooks.

These long-term outcomes are the crux of the whole issue of food additive safety. We are not looking at poisoning in the classic sense where, after ingesting a chemical, the person grabs his or her stomach and falls over. We are looking at an entire spectrum of health problems, in the long run, just as deadly. Yet the government sacrifices the long-term health issues in its approval of foreign chemicals in your foods. Why?

* * *

I have an answer and I don't like it. The answer comes out of my Ottawa schooling. As I said, I was so naive when I was first appointed to CEAC and entered the world of politicians and bureaucrats. I thought they were all there to serve the public. But having graduated from Chemical Politics 101, I now understand the motivation behind decisions about foreign chemicals in your life.

That motivation is captured in a medieval painting showing King Charlemagne at a banquet. The painting shows the King and Queen sitting at the head table and next to the King stands the official food-taster, a napkin draped over his shoulder.

Kings in those days lived in fear – and for good reason – that someone was trying to poison them. So, before a dish brought from the kitchen was placed before the King, the official food taster took a bite in the presence of the King. When I first saw this painting, I thought, oh yeah, here's the first food-safety bureaucrat.

Looking at the painting I begin to daydream, imagining myself in the position of that food taster. Am I more concerned about the King's safety or my own head? As a food taster, I'm good at detecting poisons that act quickly. If I miss the poison and the King sickens or dies within the hour, everyone knows that I failed and I'm trotted off to the executioner's block.

But – and this is a big but – if the poison doesn't work quickly and the King gets sick later, he won't blame me. He won't associate the sickness with the pork pie he ate a month ago. The doctors chalk it up to fate. I keep my head and my job. Thus I have no incentive to worry about slow-acting poisons.

Still day dreaming, I fast-forward to the present and put myself behind the pencil box of an Ottawa food-safety bureaucrat. Nothing has changed. Medieval thinking prevails. I make sure that fast-acting poisons don't get into the nation's food. People can eat food saturated with food additives and won't get sick tonight, tomorrow or the next day. Because if they did and associated the illness with the food, they'd get angry at me – the food-safety bureaucrat – for not doing a better job. But, if a woman comes down with breast cancer because of exposure to foreign chemicals when she was in her own mother's womb she'll never blame the chemicals, or more importantly, blame me or my minister.

So that's where it is today, my odyssey ended. The reasons why the government shuns modern science when it comes to safety of food-additives are laid bare. Other fields of human endeavour burst with cutting-edge ideas and technology. In contrast, the government's policy for safety testing food additives wears a hooded, medieval look because it serves the interest of politicians, bureaucrats and major food industry sectors.

Where does that leave you the consumer? Just because the government isn't looking out for your best interests does not mean you are helpless. That's why a book like *Hard to Swallow: The Truth About Food Additives* is important. It arms you with knowledge that government food-safety bureaucrats will never hand to you, namely knowledge to make an informed consumer choice. The food industry in its diversity, makes fine products as well as many that are dangerous. Which is which? This book helps you learn the answers to that question.

Acknowledgements

We want to take this opportunity to thank those who encouraged and helped us twenty years ago. For this revised edition, we sincerely thank all those people who wrote to us over the years as it kept the idea alive and encouraged us to undertake this enormous task once again.

Profound thanks to Dr. Ross Hume Hall who has been our greatest supporter. Dr. Hall's timely suggestion that we also investigate genetic engineering sent us on an odyssey which easily could have turned into a book of its own. He was extremely patient in sharing his knowledge, and his belief in us and our work often kept us going even when we became discouraged.

A word of thanks to the many authors who are providing the public with vital information concerning the food situation. In particular, we would like to thank Richard Wolfson, Elaine Gottschall and Brewster Kneen. Without their work, ours would have been much more difficult.

Many others have contributed in various ways and the following deserve special mention: Susan Nind, Jock and Jay Mullin, Kim Sarjeant, Marlene Scribner, Jeanette Perry, Rosemary Gallant and Cynthia Dearborn.

Our sincerest thanks everybody at ***alive*** **books** who contributed to the shaping of this book. It is important that we give special thanks to Lucy Kenward and our editor, Paul Razzell. Paul's enthusiasm and genuine dedication to our work was instrumental in bringing this book to completion.

There are some people who must remain nameless, especially our Angel of Mercy. You know who you are and we'd like you to know how much your advice and encouragement meant.

* * *

To my mom, Yvonne, who loved me and taught me the value of simple living.

To my beautiful children, Kim and Jonathan, whose hearts beat in mine. I hope that someday, some way, they will do their little bit for humanity.

To my special mentor who taught me that with perseverance you can climb any mountain and that the sacred tapestry of life has many threads.

To my husband, John, who never received due recognition for his encouragement and help with the first book. He has my gratitude for not allowing the household to disintegrate during the revisions for this new edition.

A word of thanks to Patricia Lockhart, a sweet and gentle bank manager. Without her, this book would have taken longer to become a reality.

Doris Sarjeant

* * *

To my husband, Harry, whose love and encouragement made it possible for me to do this work.

To my children, Jake, Jodie and Joshua who have been my inspiration for writing this book and who have had to patiently wait for it to be completed so we can get back to the business of living.

To my mom, Madelyn, who always believed in me and simply loved me.

To all my friends and family members who have supported me during this busy time with their patience and encouragement.

Karen Evans

How to Use This Book

Hard to Swallow: The Truth About Food Additives will open your eyes to the vast array of dangerous chemicals that are permitted for use in our food supply and will give you a clear understanding of some of the alarming processes used to manufacture the foods you eat every day. We guarantee you will be shocked.

The Alphabetical Guide is a list of nearly 300 food additives that have been compiled from the Canadian *Food and Drugs Act* and the Food and Agricultural Organization/World Health Organization (FAO/WHO) Committee on Food Additives. This section examines how individual food additives are used and identifies the foods in which they appear. The Alphabetical Guide also evaluates the additives' potential for harm in light of the most current scientific research. Begin by reading the lists of ingredients on the products in your kitchen cupboard, then look up individual additives in the Alphabetical Guide to learn what you are really eating – and be prepared to open your garbage can.

In the next two sections we examine two controversial processes that are changing the very nature of the foods we eat: genetic engineering and irradiation. These sections will give you a detailed overview of the issues and dangers involved in these developments in food technology and will explain why you should be concerned for the health of your family and the health of the environment.

Given that so many additives and processes can be used in the manufacture of a single food, we have presented a case study of infant formula in order to show the dangers that can be present in even the most innoccuous and supposedly "basic" foods.

The next section consists of excerpts from the Canadian *Food and Drugs Act*. This section lists all the additives that are permitted in specific foods and you will discover that there are countless additives that are not required to be listed on a product label. Without the help of this section in particular, you will never be able to determine which additives are permitted in the foods you eat every day.

We have set many passages in these regulations in bold type to draw your attention to the dangers of certain additives and to highlight areas where we believe the *Food and Drugs Act* is ineffective in protecting the public from harm. The Authors' Notes will guide you through the technical details and jargon used by Health Canada bureaucrats. The best way to use this section is to go to the Table of Contents on page 144 and look up the foods that you are most concerned about. Then turn to the sections of the Act that discuss those foods and you will quickly see that foods such as bread, margarine and cream, to name but a few, can be complex concoctions of chemical additives.

The final sections of the book deal with other legislative issues, namely Canada's field testing legislation and the laws governing the labelling of genetically engineered foods.

As you will see, your food is the focus of a very complex web of legislation, corporate interests, environmental and nutritional issues and it may be that you will never look at foods in the same way again. This is our hope. *Hard To Swallow: The Truth About Food Additives* gives you all the information you need to distinguish the wheat from the chaff and to become a well-informed and healthy consumer. Good reading!

1

Alphabetical Guide to Food Additives

Acacia Gum – *See* Gum Arabic.

Acesulfame-K (Acesulfame-potassium) – may be used in table-top sweeteners, carbonated beverages, chewing gum, breath fresheners, confectionery, some dairy beverages, desserts, fillings, mixes, fruit spreads, salad dressings, bakery mixes and bakery products as an artificial sweetener. A derivative of acetoacetic acid, acesulfame-K is 200 times sweeter than sugar and is one of the newest artificial sweeteners on the market. It was approved for use by the US Food and Drug Administration (FDA) in 1988 despite warnings from the Center for Science in the Public Interest (CSPI) that it had not been adequately tested. The CSPI claimed that tests did not meet FDA standards. Even so, these tests indicated that acesulfame-K caused lung tumours, breast tumours, several forms of leukemia and chronic respiratory disease in rats. Since acesulfame-K has a bitter aftertaste, it is usually used with sugar or aspartame to mask its bitterness. Given the adverse effects of each of these chemicals when taken singly, who knows what hazards this combination might produce? Authorized for use in Canada and the United States.

Acetic acid – may be used in cream and processed cheeses, canned asparagus, gelatin and unstandardized foods as a pH adjusting agent, an acid-reacting material and/or a water-correcting agent. In preserved fish, meat and some unstandardized foods, it is used as a preservative. Acetic acid is a colourless liquid that occurs naturally in certain foods. Food manufacturers can produce it synthetically from alcohol and acetaldehyde. Acetic acid gives substances their sharp taste and odour and is the active ingredient in vinegar. In its undiluted form, it is extremely corrosive and the bronchial tubes may become blocked if the vapours are inhaled. Acetic acid is used industrially to make oils and resins more soluble. Authorized for use in Canada and the United States.

Acetic anhydride – is used as a modifying agent in starch. It causes irritation and tissue death when it is in a gaseous state, therefore contact with skin and eyes should be avoided. The FDA suggests that modified starch be studied further. In Canada, the use of acetic anhydride is governed by the Good Manufacturing Practice regulation which means there are no regulations to limit the quantities permitted in food. Authorized for use in Canada and the United States. *See* Starch.

Acetone – may be used in spice extract, natural extractive and meat marking inks as a carrier or extraction solvent. It is also used in the manufacture of waxes, plastics, varnishes, nail polish remover, explosives and chloroform. Continued topical use may cause skin irritation. Headaches, fatigue, intoxication and irritation to the lungs may result if large amounts of acetone are inhaled. Authorized for use in Canada and the United States.

Acetone peroxide – may be used in bread, flour, whole wheat flour and unstandardized bakery foods as a bleaching, maturing and/or dough-conditioning agent. It is a strong oxidizing agent and can be harmful to the skin and eyes in a concentrated solution. Acetone peroxide is being studied for mutagenic, teratogenic, subacute and reproductive effects by the Food and Drug Administration. Canadian manufacturers are permitted to use this additive in unlimited quantities under the loose regulation known as Good Manufacturing Practice. Authorized for use in Canada and the United States.

Adipic acid – may be used in unstandardized foods as a pH adjusting agent, acid-reacting material and/or water-correcting agent and may be used in starch as a modifying agent. Limited testing has indicated that rats, and theoretically humans, assimilate adipic acid with no known adverse effects. It can also be used in the manufacture of plastics and artificial resins. Authorized for use in Canada and GRAS in the United States. *See* Starch.

Agar-agar – may be used in milk, skim milk, cream, processed cheese, relishes, ice cream, sherbet, calorie-reduced margarine, jellies, head cheese and unstandardized foods as an emulsifying, gelling, stabilizing and/or thickening agent. It is also used as a bulk laxative. Agar-agar is obtained from certain seaweeds and has been known to cause allergic reactions. Authorized for use in Canada and GRAS in the United States.

Alginate or Algin (Alginic acid, Ammonium alginate, Calcium alginate, Potassium alginate, Propylene glycol alginate, Sodium alginate) – may be used in infant formula, milk (naming the flavour), cream, beer, mustard pickles, salad dressings, sour cream, relishes, cottage cheese, processed cheese, processed cream cheese, ice cream, sherbet, calorie-reduced margarine, unstandardized foods, canned peas and asparagus, canned green beans and wax beans as an emulsifying, gelling, stabilizing and/or thickening agent. The alginates are derived from specific seaweeds. Short-term feeding studies have shown that alginates are not overtly toxic. However, another source indicated that alginates inhibited the absorption of essential nutrients in test animals. This is an area of particular concern as alginates are permitted in infant formula. Authorized for use in Canada and GRAS in the United States.

Alginic acid – *See* Alginate.

Alkanet – may be used in jams, jellies, bread, butter, margarine, cheese, milk, icing sugar, pickles, relishes, vegetable fats and oils, unstandardized foods and many other processed foods as a colouring agent. Alkanet was previously employed medically as an astringent. It is extracted from the root of the alkanet tree found in Asia Minor and the Mediterranean. In 1988, the FDA withdrew authorization for the use of this additive. Authorized for use in Canada. *See* Artificial colour.

Allura red (Red #40) – *See* Artificial colour.

Alpha amylase – *See* Amylase.

Alum – *See* Potassium Aluminum Sulphate.

Aluminum metal – may be used in bread, butter, concentrated fruit juice (unfrozen varieties), ice cream mix, icing sugar, jams and jellies with pectin, liqueurs and alcoholic cordials, milk, pickles; relishes, sherbet, skim milk, smoked fish, lobster paste and fish roe (caviar), tomato catsup, liquid, dried or frozen whole egg or egg yolk, vegetable fats and oils, margarine, a variety of cheeses and unstandardized foods as a colouring agent. The maximum level of aluminum metal permitted in the majority of cases is governed by the Good Manufacturing Practice regulation. *See* Artificial colour.

Aluminum calcium silicate – *See* Calcium aluminum silicate.

Aluminum potassium sulphate – *See* Potassium aluminum sulphate.

Aluminum sodium sulphate (Sodium aluminum sulphate) – may be used in pickles, relishes and unstandardized foods as a firming agent; in flour and whole wheat flour as a carrier for benzoyl peroxide and in baking powder as a pH adjusting agent, acid-reacting material and/or water-correcting agent. It is used medically as an astringent. In 1980, the FDA decided to continue its GRAS status under the Good Manufacturing Practice regulation. Also authorized for use in Canada.

Aluminum sulphate – may be used in canned crabmeat, lobster, salmon, shrimp, tuna, pickles, relishes and unstandardized foods as a firming agent; in liquid or frozen whole egg, egg white (albumen) or egg yolk to stabilize the albumen during pasteurization and is used in starch as a modifying agent. Aluminum sulphate is also used to tan leather, size paper, purify water, treat sewage, waterproof concrete and to deodorize and decolour petroleum. It is also a component of antiperspirants and can be used as a medical disinfectant. Authorized for use in Canada and GRAS in the United States. *See* Starch.

Amaranth (Red # 2) – *See* Artificial colour.

Aminoacetic acid – *See* Glycine.

Ammoniated glycyrrhizin (Monoammonium glycyrrhizin) – is not considered an additive according to Division 1 of the *Food and Drugs Act*, B.01.009.(2). It is the main flavour in licorice. It can also be used in baked goods, candy and beverages such as root beer. Ammoniated glycyrrhizin is a powerful drug that can cause heart failure, elevated blood pressure, swelling and headaches. This chemical was on the FDA's top priority list to be tested for its ability to cause cancer and birth defects. Authorized for use in Canada and GRAS in the United States.

Ammonium alginate – *See* Alginate.

Ammonium bicarbonate (Ammonium carbonate) – may be used in chocolate, cocoa, milk chocolate, sweet chocolate and unstandardized foods as a pH adjusting agent, acid-reacting material and/or water-correcting agent. It is an alkali with a slight odour of ammonia, which is produced by passing carbon dioxide gas through concentrated ammonia water. Ammonium bicarbonate is also used medically in soothing baths, as an expectorant and to relieve gas. Authorized for use in Canada and GRAS in the United States.

Ammonium carbonate – *See* Ammonium bicarbonate.

Ammonium carrageenan – *See* Carrageenan.

Ammonium chloride – is used in flour, whole wheat flour, bread and unstandardized foods as a yeast food. Ammonium chloride can be used industrially in batteries, dyes, electroplating and safety explosives. It is used medically as an expectorant, a diuretic and a treatment to make urine more acidic. Ammonium chloride may cause nausea and vomiting, and as with all concentrated solutions of ammonia compounds, it can cause skin irritations. In 1980, the FDA decided to extend the GRAS status of ammonium chloride for "packaging only" under the Good Manufacturing Practice regulation. Also authorized for use in Canada.

Ammonium furcelleran – *See* Furcelleran.

Ammonium hydroxide – may be used in chocolate, cocoa, milk chocolate, sweet chocolate, gelatin and unstandardized foods as a pH adjusting agent, acid-reacting material and/or water-correcting agent. It is an alkali which exists only in water solutions and is formed by dissolving ammonia gas. As with all ammonia compounds, concentrated solutions can cause skin irritations. Authorized for use in Canada and GRAS in the United States. *See* Caramel.

Ammonium phosphate, dibasic (Diammonium phosphate) – may be used in ale, bacterial cultures, baking powder, beer, light beer, malt liquor, porter, stout and unstandardized bakery foods as a pH adjusting agent, acid-reacting material and/or water-correcting agent. It is also used in bread, cider, honey wine, wine and unstandardized bakery foods as a yeast food. It is employed industrially to fireproof textiles, paper and wood and to prevent afterglow in matches. It is also utilized in dentrifices, corrosion inhibitors and in fertilizers. Medically, it is used to reduce body water and to make urine more acidic. Authorized for use in Canada and GRAS in the United States. *See* Phosphoric Acid.

Ammonium phosphate, monobasic (Monoammonium phosphate) – may be used in ale, bacterial culture, baking powder, beer, light beer, malt liquor, porter, stout, unstandardized bakery foods and uncultured buttermilk as a pH adjusting agent, acid-reacting material and/or water-correcting agent; and in bread, ale, beer, cider, honey wine, light beer, malt liquor, porter, stout, wine and unstandardized bakery foods as a yeast food. It can be employed industrially to fireproof certain materials, and is also used in dentrifices and fertilizers. Medically, it is used to reduce body water and to make urine more acidic. Authorized for use in Canada and GRAS in the United States. *See* Phosphoric Acid.

Amylase – may be used in, bread, flour, fruit juice, instant cereals, ale, beer, malt liquor, porter, stout, cider, wine, chocolate syrup and unstandardized bakery products as a food enzyme. It can be prepared from hog pancreas and is used to change starch into sugar. Amylase occurs naturally in plant seeds and is present in saliva, pancreatic juices and micro-organisms. It is produced commercially from barley malt and the fungi *Aspergillus niger* and *Aspergillus oryzae*. The *Food and Drugs Act* lists eleven different sources from which amylase can be taken, four of which are genetically altered. This frightening new trend is covered in more depth in Genetically Engineered Foods, page 109. Authorized for use in Canada and the United States.

Anise – is not considered to be an additive according to Division 1 of the *Food and Drugs Act*. It is a fruit found in Asia, Europe and the United States. Anise is used by the food industry principally as a seasoning and a flavouring in various products and has a taste similar to licorice. Although this food additive appears to be non-toxic, information on test results is very limited. Authorized for use in Canada and GRAS in the United States.

Annatto – may be used in jams, jellies, bread, butter, margarine, cheese, milk, icing sugar, pickles, relishes, vegetable fats and oils, unstandardized foods and many other processed foods as a colouring agent. Authorized for use in Canada and the United States. *See* Artificial colour.

Arabinogalactan (Larch gum) – may be used in essential oils, unstandardized dressings, pudding mixes, beverage bases and mixes and pie filling mixes as an emulsifying, gelling, stabilizing and/or thickening agent in levels consistent with Good Manufacturing Practice. It is extracted from the wood chips cut from the larch gum tree which is found in abundance in the Northwestern United States. Arabinogalactan has been found to contain large amounts of tannic acid and other impurities. At present, it is the tannic acid content that makes this additive suspicious as scientists think it may be a weak carcinogen. It is on the FDA's priority list of additives to be studied for mutagenic, teratogenic, subacute and reproductive effects. Authorized for use in Canada and the United States.

Arsenic – is not considered a food additive but residual limits are permitted in certain levels in fruit juice, fruit nectar, ready-to-serve beverages, fish protein, edible bone meal and water other than mineral water or spring water that is sold in sealed containers. *See* Table I, Division 15, Regulation B.15.001 in the *Food and Drugs Act*. According to the forty-fourth report of the FAO/WHO Expert Committee on Food Additives, 1995, the need for limits for arsenic in specific foods will be assessed on a case-by-case basis in the future. Such limits will be reduced or withdrawn unless the information provided indicates that limits for arsenic are necessary. Authorized for use in Canada and the United States.

Artificial colour – is a blanket term used to describe any one of a number of colours which the government permits in foods. Food manufacturers are not required to list the chemicals used to produce any of the following products:

- food colour preparations
- flavouring preparations
- artificial flavouring preparations
- spice mixtures
- seasoning or herb mixtures
- vitamin preparations
- mineral preparations
- food additive preparations
- rennet preparations
- food flavour-enhancer preparations
- compressed, dry, active or instant yeast preparations

According to Division 1, B.01.010 (b) a manufacturer can use a number of chemicals to produce a specific colour and then use a combination of separate colours in its product and list them on the label merely as colour(s) or artificial colour(s).

Dr. Ross Hume Hall, former head of the Department of Biochemistry at McMaster University echoes what many other prominent scientists are saying: "What possible benefit do these dyes confer on the consumer? The garish colours just mask poor quality foods. If a food has to be dyed, there is something wrong with it." Colours can replace the more expensive natural ingredients in some products, such as certain fruit drinks, which are made up almost entirely of chemicals. With the addition of chemicals, food manufacturers can create a vivid look for their highly-processed foods which otherwise would look washed out. Remove the dyes and chemicals and what is left would not resemble food.

Artificial colour is one of the most hazardous mixtures of chemicals that is added to food. A large percentage of these chemicals, whether natural or synthetic, have never seen the inside of a laboratory for testing for adverse effects and yet they are being ingested every day by almost the entire North American population.

Coal tar dyes, the largest class of dyes used in foods, are also used to colour fabrics, cosmetics, paints and inks. The carcinogenicity of many coal tar dyes is well documented. Unfortunately for consumers, it is nearly impossible to know which chemicals are in the products they are buying since Canadian legislation prescribes that only the word "colour" need be listed on a label. In the late 1970s, the Canadian government responded to growing consumer concern over the use of artificial colours by saying: "Food additives such as colour and texture agents are used in Canada to correct natural variations in foods so that a particular food looks, tastes and performs the same way every time." Surely, if consumers understood that natural variations occur in foods due to seasonal changes, for example, they would choose the natural product over ones that have been cosmetically repaired with colours and flavours.

A quick survey in any grocery store will confirm for you that most products contain added colour. According to Dr. Ross Hume Hall, "The food industry, aware of the public's aversion to the word artificial, in the 1970s, began to research new colours and flavours made from natural sources – as opposed to coal tar. They now have a range of such colours and flavours that, like coal tar products, can be mixed to get any shade and flavour desired. Thus, you might buy a raspberry sorbet which lists as ingredients natural colours and flavours. You might think isn't that nice, they put in some additional raspberry flavour and colour made from raspberries. Not a chance. The flavours and colours come from sources that have never been near a raspberry."

Prominent researchers have been warning us for years of the dangers of food colours. In the words of Dr. Hall, "This business of taking separate chemicals and going through tortuous experiments with each is not an adequate approach. They should look at the whole class of chemicals and how they are made."

An investigation by independent researchers in Canada in the 1970s confirmed the connection between food and aggressive behaviours. Dr. J. Ivan Williams, of the University of Toronto, noted that the behaviour of hyperactive children generally deteriorated after eating foods that contained synthetic colour additives. Hyperactivity, or hyperkinesis, is characterized by restlessness, excitability, short attention span, impatience, aggressiveness, impulsiveness and disruptive classroom behaviour.

Studies have shown that children's behaviour improves when they eat nutritious foods, particularly those without artificial colours. Conversely, behaviour deteriorates when they are fed processed foods containing artificial colour. These effects are most pronounced in younger children in the three hours after the food was eaten.

Dr. Ben Feingold, who is considered a pioneer in the field, asserts that forty per cent of so-called hyperkinetic or hyperactive children are really suffering from a sensitivity to certain natural food components such as salicylates and synthetic food additives, specifically artificial colours and flavours. By eliminating these elements from the diets of the children he treated, he was able to discontinue medication and improve the behaviour of more than fifty per cent of them. There was also a dramatic improvement in their performance at school.

The number one mental disease in North America is schizophrenia (meaning split mind). It has been estimated that people suffering from schizophrenia fill sixty per cent of the beds in mental institutions and the cost of treating them is an astonishing $15 billion annually. Dr. Lesser, a psychiatrist at the University of California in Berkeley, treats schizophrenics with what he considers a "new wonder drug" which consists of a nutritious diet including whole grains, fresh fruits and vegetables and plenty of high-quality protein. As schizophrenics tend to consume large quantities of junk foods, Dr. Lesser's treatment brings about a dramatic improvement in these patients.

An ever-increasing number of doctors, psychiatrists and medical researchers are realizing that a diet of food that is free of chemicals and mechanical degradation is an effective cure for diseases and disorders of the body and mind. The medical profession, as a whole, remains skeptical, however, and this attitude is understandable given that very little time is devoted in medical school to the study of nutrition or the effects of eating chemical food additives. This leaves doctors uninformed on the subject.

The following is a list of food colour additives. Along with the colours that are still permitted for use in food, we have included any we could locate that have been banned. We included the banned colours to stress our point that all synthetic food dyes, especially the coal tar dyes, are inherently dangerous. A look at the synthetic colour additives listed below indicates the majority of them have been banned in many countries due to their toxic effects. In fact, in 1979, Norway banned the use of all colour additives in food. Canada and the US have refused to follow suit, claiming that they have nothing with which to replace those specific additives. If the use of all colours were banned, the number of artificial foods or highly processed foods would be drastically lowered. Nobody would buy these over-processed, washed out looking products. People would go back to growing more of their own or buying from local organic farmers and rediscover what real food is supposed to taste and look like.

Since Health Canada has a dismal record when it comes to safeguarding the food supply, consumers must take responsibility for protecting themselves and their children. Let the buyer beware. When consumers demand changes and refuse to purchase certain products, food manufacturers respond – without instruction from government. With this in mind, it obviously will be more effective to go directly to the source – the food manufacturer. As a matter of fact, one manufacturer told us, "If you are not happy with my product, call me. After all, if consumers stop buying my product, I'm out of business and that's not good for me." If manufacturers get enough calls from consumers saying that they are not going to buy their products because of all the dangerous additives, they will remove them because they want to keep your business. Also, call or write your elected official. Public pressure may encourage them to become pro-consumer. It is time that Canadians and Americans took a stand along with countries such as Norway.

Allura red (Red #40) – is permitted for use in bread, butter, ice cream mix, ice milk mix, milk, jams and jellies with pectin, pickles, relishes, icing sugar, liqueurs and alcoholic cordials, sherbet, skim milk, smoked fish, lobster paste, caviar, tomato catsup, prepared fish and meat, salted anchovy, salted shrimp and unstandardized foods.

This dye was originally developed to replace red dye #2 and red dye #4 which were banned in 1976. The FDA approved red dye #40 at that time, even though the testing proved deficient. The experiments did not include two species of animals as stipulated by the guidelines for the testing of food additives. The British and Canadian government would not approve red dye #40 unless additional tests were done. In 1976, Allied Chemical completed these tests and the results indicated the dye caused cancer in mice. A United States government pathologist at the time was quoted as saying that this additive had all the properties of a carcinogen. It is banned in Sweden, Norway, Japan, Italy, Israel, France, Austria, the United Kingdom and Australia. Authorized for use in Canada and the United States.

Amaranth (Red #2) – is permitted for use in bread, butter, ice cream mix, ice milk mix, milk, jams and jellies with pectin, pickles, relishes, icing sugar, liqueurs and alcoholic cordials, sherbet, skim milk, smoked fish, lobster paste, caviar, tomato catsup, prepared fish and meat, salted anchovy, salted shrimp and unstandardized foods. In 1971, a Soviet study found that this dye was associated with birth defects and interfered with reproduction in rats and even induced cancer. In 1975, the FDA conducted new tests that confirmed that the dye increased the number of malignant tumours in female rats. Other studies have shown that amaranth (red #2) prevented pregnancies and caused stillbirths in rats which were injected with amaranth. The FDA finally banned the dye in January 1976. Norway banned the use of all synthetic food colouring in 1979. It has also been banned in Austria, Finland, France and Italy (except for use in caviar), Greece, Japan, Yugoslavia and the former Soviet Union. Authorized for use in Canada.

Blue #1 – *See* Brilliant Blue.

Blue #2 – *See* Indigotine.

Brilliant blue (Blue #1) – is a coal tar derivative permitted for use in bread, butter, ice cream mix, ice milk mix, milk, jams and jellies with pectin, pickles, relishes, icing sugar, liqueurs and alcoholic cordials, sherbet, skim milk, smoked fish, lobster paste, caviar, tomato catsup, prepared fish and meat, salted anchovy, salted shrimp and unstandardized foods.

This synthetic food colouring was the subject of a number of studies and investigations in the US and elsewhere. The studies revealed that brilliant blue caused cancer when ingested, as well as malignant tumours in rats at the sites of injection. Other studies indicated this colouring promoted breast tumours in test animals and increased the incidence of kidney tumours in male mice. The FAO/WHO recommended that it not be used. This dye has been banned in all European Commonwealth Countries as well as in Austria, Finland, Norway, Sweden and Switzerland. Authorized for use in Canada and the United States.

Butter yellow – banned in 1919.

Canthaxanthin – may be used in bread, butter, concentrated fruit juice (unfrozen varieties), ice cream mix, icing sugar, jam with pectin, liqueurs, milk, pickles, relishes, sherbet, smoked fish, lobster paste, caviar, tomato catsup; liquid, dried or frozen whole egg, vegetable fats and oils, margarine, a variety of cheeses and unstandardized foods. This additive can be derived from edible mushrooms, trout, salmon and crustaceans and can be used to give a pink colour to foods. It can also be used in chicken feed to enhance the yellow colour of the birds' skin. A report from the FAO/WHO Expert Committee on Food Additives suggests that canthaxanthin may cause harm to the liver, as long-term studies on rats showed that it was toxic to their livers. It also caused crystals to deposit on the human retina – a condition that may not be reversible. Authorized for use in Canada and GRAS in the United States.

Carbon black – may be used in bread, butter, concentrated fruit juice (unfrozen varieties), ice cream mix, icing sugar, jam with pectin, liqueurs, milk, pickles, relishes, sherbet, smoked fish, lobster paste, caviar, tomato catsup; liquid, dried or frozen whole egg, vegetable fats and oils, margarine, a variety of cheeses and unstandardized foods with Good Manufacturing Practice regulating the maximum level of use in the majority of cases. At their thirty-seventh meeting, the FAO/WHO Expert Committee on Food Additives decided to change the name of carbon black to vegetable carbon. Authorized for use in Canada and the United States.

Charcoal – may be used in bread, butter, concentrated fruit juice (unfrozen varieties), ice cream mix, icing sugar, jam with pectin, liqueurs, milk, pickles, relishes, sherbet, smoked fish, lobster paste, caviar, tomato catsup, liquid, dried or frozen whole egg, vegetable fats and oils, margarine, a variety of cheeses and unstandardized foods. The maximum level of use in most of these foods is governed by the Good Manufacturing Practice regulation. This dye has been shown to contain cancer-causing by-products. It was banned in Norway in 1979 and in the US in 1976. Authorized for use in Canada.

Citrus red #2 – was first used under the name red #32. The relevant history of this dye is as follows:

- In 1956, it was banned. Subsequently, the name changed to citrus red #2 and its use continued in certain foods.

- In 1965, the University of Otega Medical School reported significant levels of bladder cancer in rodents who were fed the dye for up to twenty-four months.
- In 1969, the FAO/WHO Expert Committee issued warnings that it should not be used as a food colour.
- In 1973, researchers proved that citrus red #2 was toxic. Various studies linked it with the damage of internal organs and the production of cancer in animals. The International Agency for Research on Cancer concluded that citrus red #2 is a carcinogen in rats and mice. In the same year, it was banned from use in all edible food but was allowed to colour orange skins. In this process, oranges are heated and either sprayed with the dye or immersed in containers of hot dye. Since orange skins are used in making marmalade and some people still peel oranges with their teeth, this colour is still being ingested. At its thirteenth meeting, the FAO/WHO warned that this substance should not be used as a food colour. This additive was banned in Australia, Britain and Norway.
- In 1993, the FDA recommended a ban but to our knowledge it is still permitted for use on oranges in the United States. Authorized for use in Canada.

Erythrosine (Red #3) – is a coal tar derivative permitted for use in bread, butter, ice cream mix, ice milk mix, milk, jams and jellies with pectin, pickles, relishes, icing sugar, liqueurs, sherbet, smoked fish, lobster paste, caviar, tomato catsup, prepared fish and meat, salted anchovy, salted shrimp and unstandardized foods.

Studies on animals have verified that this dye is a carcinogen. Furthermore, erythrosine may hinder the transmission of nerve impulses in the brain by interfering with the uptake (absorption) of dopamine, a neurotransmitter. If this were the case in humans, it could cause learning disabilities in children. Tests conducted by the food colour industry in 1982 showed this dye caused thyroid tumours in male rats. A further concern was that this dye may be linked to gene mutation.

In 1983, the US National Toxicology Program, a government agency which evaluates cancer tests, confirmed that evidence was available to demonstrate that erythrosine (red #3) is a cancer-causing agent in animals. Several months later, the

FDA officially admitted that this coal tar dye is a cancer-causing agent. Business interests fought tenaciously to prevent a ban of the dye and what followed was more reviews and many more delays. Business interests aligned themselves with bureaucrats in the White House whose sympathy lay, for various reasons, with industry rather than the consumer. The following is an abbreviated example of what can take place and how long action can be delayed when the food industry goes head to head with federal laws and public-minded scientists.

- In 1979, the American National Health Institute claimed that erythrosine (red #3) may interfere with the transmission of nerve impulses to the brain.
- In 1982, the food colour industry's tests indicated the dye caused thyroid tumours in male rats. They tried to play down the results.
- In 1983, the National Toxicology Program, a government agency that specializes in evaluating cancer tests, concluded the dye caused cancer.
- In November 1983, the FDA officially declared erythrosine (red #3) a carcinogen.
- In 1983, the food colour industry mobilized to fight any ban on the dye. It aligned itself with like-minded bureaucrats in Reagan's administration. Cherry growers and producers of canned fruit cocktail lobbied for more time to test – and received it.
- In March 1984, the Acting Commissioner of the FDA urged a ban on this dye which had been proven to cause cancer. The FDA was in danger of losing its credibility as a regulatory agency, yet nothing happened.
- In October 1984, the subcommittee of the House Committee on Government Operations conducted an investigation to find out why the FDA was not banning a known cancer-causing dye. The subcommittee stated that the Department of Health and Human Services was violating the federal safety laws.
- In late 1984, legislation was introduced to ban erythrosine (red #3) and a number of other dyes suspected of being carcinogenic. The Department of Agriculture and US Senators aligned themselves with the food dye industry.

- In June 1985, the FDA granted manufacturers anywhere from one to six years to prove the dye was not a cancer-causing agent. Many right-minded officials were demoralized.
- July 1985 to 1990. The Reagan Administration's legacy to the American people was to leave the FDA laws with no teeth. The furor gradually died down and the food industry continued to produce and add cancer-causing dyes to the food supply for many years before overwhelming evidence forced the government to ban this hazardous chemical.
- In 1990, after eleven years of debate, erythrosine (red dye #3) was banned in the US with the condition that the dye remain on the market until supplies were used up. Disgraceful as it sounds, it is still authorized for use in Canada.

Fast green – *See* Green #3.

Gold – This synthetic dye is permitted in alcoholic beverages at levels consistent with the Good Manufacturing Practice regulation. Authorized for use in Canada.

Green #1 (Guinea green B) – is a synthetic dye that has been rated "E" by the WHO which means it has been found to be injurious and should not be used in food. Banned in Norway in 1979 and in the US in 1965. Not presently being used in Canada.

Green #2 (Light green SF yellow) – when injected under the skin of rats this colouring agent produced tumours at the site. Banned in Norway in 1979. Delisted in the US in 1966. It is not being used in Canada at present.

Green #3 (Fast green) – is permitted for use in bread, butter, ice cream mix, ice milk mix, milk, jams and jellies with pectin, pickles, relishes, icing sugar, liqueurs, sherbet, smoked fish, lobster paste, caviar, tomato catsup, prepared fish and meat, salted anchovy, salted shrimp and unstandardized foods. Studies in the US and elsewhere have shown that this chemical induces cancer when injected into rats. Norway, Australia, Britain and several European countries have banned this dye. Authorized for use in Canada and the United States.

Guinea green B – *See* Green #1.

Indigotine (Blue #2) – is a coal tar derivative permitted for use in bread, butter, ice cream mix, ice milk mix, milk, jams and jellies with pectin, pickles, relishes, icing sugar, liqueurs, sherbet, smoked fish, lobster paste, caviar, tomato catsup, prepared fish and meat, salted anchovy, salted shrimp and unstandardized foods.

This synthetic dye has been in use since the 1950s. A study in 1981 showed that large amounts of blue #2 increased the incidence of malignant brain gliomas, mammary cancers and bladder tumours. It produced malignant tumours where it was injected under the skin of rats. A review of this study by cancer expert Dr. Samuel Epstein confirmed the results that indigotine (blue #2) is a cancer-causing agent. The Public Citizen Health Research Group called for the immediate ban of this dye by the Food and Drug Administration. Banned in Norway in 1979. Authorized for use in Canada and the United States.

Iron oxide (Rust) – may be used in bread, butter, concentrated fruit juice (unfrozen varieties), ice cream mix, icing sugar, jam with pectin, liqueurs, milk, pickles, relishes, sherbet, smoked fish, lobster paste, caviar, tomato catsup, liquid, dried or frozen whole egg, vegetable fats and oils, margarine, a variety of cheeses and unstandardized foods.

Iron oxide is iron combined with oxygen, more commonly known as rust. It is considered a natural dye. In some of the foods mentioned above, the government places restrictions on the amount of iron oxide manufacturers may use. At the same time, they allow unlimited amounts in the remainder of these foods under the Good Manufacturing Practice regulation. One must question the rationale of restricting the amount of an additive used in a few products while allowing its unlimited use in other foods. Due to lack of testing, the effects of iron oxide on the human body are not clear. Nevertheless, it is banned throughout the European Community. Authorized for use in Canada and permitted for use in the US for dyeing eggshells and pet food.

Light green SF yellow – *See* Green #2.

Orange B – was banned in the US in 1978. As of 1985, the ban had not been finalized and any company could begin using it again. Apparently orange B was never well tested. The Center for Science in the Public Interest asked the FDA to finalize the ban as it was found to have significant similarities to the dyes red #2 and yellow #5. They had been banned because they increased the risk of cancer. Orange B was banned in Norway in 1979. It is not authorized for use in Canada.

Orange #1 – was banned in the US in 1956 and in Norway in 1979. It is not authorized for use in Canada.

Orange #2 – was banned in the US in 1960 and in Norway in 1979. it is not authorized for use in Canada.

Ponceau SX (Red #4) – may be used in or upon fruit peel, glacé fruits and maraschino cherries. The FDA banned the use of this dye in food in 1964. Dr. Kent Davis of the FDA reported that this dye caused the adrenal glands in dogs to atrophy and caused the formation of urinary bladder polyps. The cherry growers industry challenged these results on the grounds that the amount of the dye ingested by eating maraschino cherries or glacé fruits was insignificant. They won a provisional delay so other studies could be conducted. These studies were never done and in 1976 the FDA banned ponceau SX (red #4) from all foods. It is also banned in Norway and Japan. Authorized for use in Canada.

Red #1 – banned in Norway in 1979 and in the US in 1961. It is not authorized for use in Canada.

Red #2 – *See* Amaranth.

Red #3 – *See* Erythrosine.

Red #4 – *See* Ponceau SX.

Red #32 – *See* Citrus red #2.

Red #40 – *See* Allura red.

Rust – *See* Iron oxide.

Silver metal – may be used in bread, butter, concentrated fruit juice (unfrozen varieties), ice cream mix, icing sugar, jam with pectin, liqueurs, milk, pickles, relishes, sherbet, smoked fish, lobster paste, caviar, tomato catsup, liquid, dried or frozen whole egg, vegetable fats and oils, margarine, a variety of cheeses and unstandardized foods. Authorized for use in Canada.

Sunset yellow (Yellow #6) – is permitted for use in bread, butter, ice cream mix, ice milk mix, milk, jams and jellies with pectin, pickles, relishes, icing sugar, liqueurs, sherbet, smoked fish, lobster paste, caviar, tomato catsup, prepared fish and meat, salted anchovy, salted shrimp and unstandardized foods. This synthetic dye is a family member of the azo dyes, a number of which have been proven hazardous. Various studies indicate that sunset yellow causes carcinogenic activity in the kidneys and the adrenal glands of animals. The FDA suggests this could be the result of up to half a dozen carcinogenic contaminants that are present in batches of commercial dye. It has also been suggested that sunset yellow could possibly cause chromosomal damage. When one considers the history of dyes, wouldn't it be worth considering purchasing uncoloured food products? Then, perhaps, take the time to send a letter to your local government representative to have all food dyes banned as they did in Norway in 1979. This dye has also been banned in Finland and Sweden. Authorized for use in Canada and the United States.

Tartrazine (Yellow #5) – is permitted for use in bread, butter, ice cream mix, ice milk mix, milk, jams and jellies with pectin, pickles, relishes, icing sugar, liqueurs and alcoholic cordials, sherbet, skim milk, smoked fish, lobster paste, caviar, tomato catsup, prepared fish and meat, salted anchovy, salted shrimp and unstandardized foods. Tartrazine is a synthetic dye derived from coal tar. It is also a member of the chemical family known as azo dyes, a number of which are hazardous. In many people it causes allergic reactions such as runny noses, asthma and hives. It seems to pose the greatest risk to aspirin-sensitive people and has been reported to cause life-threatening asthma symptoms in this portion of the population. Banned in Austria, Finland, Sweden and Norway. Officials have recommended that it be banned throughout the European Community. Authorized for use in Canada and the United States.

Titanium dioxide - may be used in bread, butter, concentrated fruit juice (unfrozen varieties), ice cream mix, icing sugar, jam with pectin, liqueurs, milk, pickles, relishes, sherbet, smoked fish, lobster paste, caviar, tomato catsup, liquid, dried or frozen whole egg, vegetable fats and oils, margarine, a variety of cheeses and unstandardized foods. It is found naturally in certain minerals and can also be produced commercially in various ways. Lung damage can occur if large quantities of the dust are inhaled. The European Community has banned its use. Authorized for use in Canada and the United States.

Violet #1 – was originally certified for Allied Chemical around 1951 to colour Easter eggs and some candies. Canadian studies completed in 1962 indicated that the dye caused cancer but for various reasons the results were ignored. In 1973, a Japanese study proved indisputably that violet #1 caused various forms of cancer. The FDA banned the dye in 1973. We were unable to find evidence of its use in Canada.

Yellow #1 – banned in 1959 in the US and in Norway in 1979. We were unable to find evidence of its use in Canada.

Yellow #2 – banned in the US in 1959 and in Norway in 1979. We were unable to find evidence of its use in Canada.

Yellow #3 – banned in the US in 1959 and in Norway in 1979. We were unable to find evidence of its use in Canada.

Yellow #4 – banned in the US in 1959 and in Norway in 1979. We were unable to find evidence of its use in Canada.

Yellow #5 – *See* Tartrazine.

Yellow #6 – *See* Sunset yellow.

Artificial flavour – is a blanket term used to describe any of the numerous flavouring preparations which are added to almost all foods. Flavours are exempt from the definition of food additives; *see* Division 1, Section B.01.001 of the *Food and Drugs Act.*

Some of the functions performed by flavours are

- to mask poor quality foods
- to replace the more expensive natural ingredients in some products
- to restore the original taste of highly processed foods

There has been very little research done on the estimated two thousand chemicals which can be used in a variety of combinations to produce specific flavours.

Division 1, Section B.01.010 further states that one or more flavours or artificial flavours may be shown on a list of ingredients by the common name "flavour" or "artificial flavour" respectively. This means that a manufacturer can use almost any number of chemicals to produce a specific flavour and incorporate a combination of these flavours in its product and label them as "flavour(s)" or "artificial flavour(s)."

The FDA has taken a similar approach, by making it unnecessary to list specific flavouring ingredients on food labels.

Our understanding of these regulations is that a food manufacturer could produce a new flavouring and with little or no testing call it "safe" and add it to foods without acquiring permission from the government. The fact that flavour manufacturers periodically submit to the government a list of new ingredients which they are adding to foods confirms our understanding of these regulations. In recent years the use of the word "natural" on labels has become more and more common. Consumers may believe this "natural" product to be more wholesome. However, we need to ask ourselves "when is 'natural' really natural?" The fact that food manufacturers can use any number of chemicals (some of which may be derived from nature) to produce a specific flavour, then use any combination of these flavours in their product and list it on the label as "flavour," "artificial flavour" and possibly "natural flavour" is reason to question the term "natural." Usually, if you see one of these terms on a label you can be almost certain that the product contains little or none of the ingredients from which that flavour would normally be derived. For interest's sake, we have included a recipe for an imitation cherry flavour taken from *The Complete Eater's Digest and Nutrition Scoreboard* by Michael F. Jacobson. Note it is a combination of thirteen chemicals:

eugenol
cinnamic aldehyde
anisyl acetate
anisic aldehyde
ethyl oenanthate
benzyl acetate
vanillin
aldehyde C^{16} (strawberry aldehyde)
ethyl butyrate
amyl butyrate
tolyl aldehyde
benzaldehyde (primary flavour)
alcohol-95% (solvent)

Government and industry defend these lenient regulations by saying that flavourings are used in relatively small amounts and that almost all of them occur naturally. Some researchers feel that just because a chemical is used in minute amounts doesn't guarantee its safety. In fact, these small amounts may be even more dangerous as the body may not recognize them and either store them or send them directly to the bloodstream – the potential for harm to the human body is cause for serious concern. We all recognize by now that not all naturally occurring chemicals are safe. Furthermore, nobody is sure of the results of combining several chemicals, natural or otherwise. The following is a list of a few flavourings that have been tested.

- Allye isothiocyanate is found naturally in oil of mustard. It was reported to cause cancer in male rats.
- Benzyl acetate can be chemically manufactured and is also found in some flowers. It was found to cause pancreatic cancer in rats and liver tumours in mice.
- Cinnamyl anthranilate is produced synthetically. It was found to cause liver cancer in mice and pancreatic and kidney cancer in rats. The FDA proposed a ban of this flavouring in 1982 but to our knowledge the ban has never been finalized. Although this flavouring is no longer in use, the fact that the ban has not been finalized means it could be used again.
- Oil of calamus is a natural flavour that has been proven to cause tumours to grow in the intestines of laboratory animals.
- Quinine is a flavouring used in tonic water. It has been suggested that pregnant women should avoid quinine as it may have toxic effects on the fetus. Testing has indicated that hearing may be impaired from eating large amounts of quinine. As a point of interest, a pilot informed us that the Regional Medical Health Examiner of Transport Canada had warned pilots that the quinine in tonic water "collects in the middle ear over a period of time which can throw off the equilibrium" and should therefore be avoided.
- Safrole occurs naturally in sassafras. It was found to cause cancer of the liver.

Dr. Ben F. Feingold, author of *Why Your Child is Hyperactive*, has suggested that artificial flavours and colours used in processed foods and beverages are a cause of allergies and behavioural problems in children. When these substances were eliminated from the diet of people he treated, their condition improved rapidly. Dr. Feingold has linked the sharp increase in hyperactivity and other learning disabilities to the escalating sales of artificially flavoured and coloured foods and beverages in recent years. Since artificial flavours and colours make up about eighty per cent of all food additives, these discoveries are particularly significant.

Of all the allergic responses attributed to food additives, more have been linked to artificial flavourings than to any other group of additives. Moreover, due to the widespread use of these substances, almost everyone has contact with them daily. Some of the reported disorders associated with flavours fall under the following headings: skin disorders, respiratory problems, blood abnormalities, skeletal ailments, gastrointestinal upsets and neurological disturbances such as behavioural disorders and migraine headaches.

Several of the artificial flavours permitted for use are suspected carcinogens (cancer-causing agents). The chief of the Environmental Cancer Section of the US National Cancer Institute recommended that approximately three hundred flavouring agents should be studied in light of the evidence that they cause cancers in rats. Given this evidence, consumers are now confronted with a problem of major significance.

Ascorbic acid (Vitamin C) – may be used in flour, whole wheat flour, bread and unstandardized bakery foods as a bleaching, maturing or dough-conditioning agent; in beer, canned mushrooms, apple sauce, peaches and tuna, cider, frozen fruit, head cheese, preserved fish, poultry meat, wine, fats and oils, lard, mono- and diglycerides, shortening and unstandardized foods as a preservative. Natural vitamin C is necessary for healthy bones, teeth and blood vessels. This vitamin deteriorates quickly if exposed to air when in solution form. The ascorbic acid used in our food is produced synthetically and is considered to be non-toxic. Authorized for use in Canada and GRAS in the United States.

Ascorbic acid calcium salt – *See* Calcium ascorbate.

Ascorbyl palmitate – is used in fats and oils, mono- and diglycerides, lard, shortening, margarine and unstandardized foods as a Class IV preservative. It is produced by combining vitamin C with palmitic acid. It is considered to be non-toxic. Authorized for use in Canada and GRAS in the United States.

Aspartame (NutraSweet™, Equal™) – is a white, odourless, crystalline powder 180–200 times sweeter than sugar. It may be used in or upon table-top sweeteners at levels consistent with the Good Manufacturing Practice regulation which means it can be used in any amount needed to achieve the desired effect. It is popularly used in diet soft drinks and diet foods and is also used in breakfast cereals, beverages, desserts, toppings, fillings, chewing gum, breath freshener products; fruit spreads, purées and sauces, table syrups, salad dressings, peanut spreads, condiments, coating mixes for snack foods and confections. It can also be used in bakery products and baking mixes in encapsulated form to prevent degradation during baking.

The controversy surrounding aspartame has raged for nearly thirty years, ever since the G.D. Searle company first approached the FDA in the late 1960s to request permission for its use. G.D. Searle is now a division of Monsanto. To unravel the history of this controversial additive, we will examine the composition of aspartame, look at its track record and follow its progress through the halls of Washington.

The Consumer Safety Network, a non-profit organization, claims that approximately ten thousand complaints about aspartame have been reported. Some of the reported adverse effects are headaches, nausea, vertigo, insomnia, numbness, blurred vision, memory loss, depression, personality changes, hyperactivity, seizures, rashes, anxiety attacks, muscle cramping and joint pain, loss of energy and hearing loss. In light of the debilitating symptoms which aspartame has caused in large numbers of people, it is now also being considered for possibily triggering chronic fatigue syndrome, Alzheimer's disease, post polio syndrome, epilepsy, anxiety/phobia disorders, manic depression, Graves' disease and multiple sclerosis.

The three components of aspartame are aspartic acid (40%), methanol (10%) and phenylalanine (50%). When a diet drink is held at temperatures over 29.4°C (85°F) for a few weeks, there is no aspartame left in the soft drinks, just the components it breaks down into, like formaldehyde, formic acid and diketopiperazine, a chemical which can cause brain tumours. All of these substances are known to be toxic to humans.

Aspartic acid – One of the main components of aspartame is aspartic acid. Aspartic acid is an amino acid often referred to as an excitotoxin which is an agent that can cause sensitive neurons to die and can even cause brain lesions. Dr. Russell Blaylock, in *Excitotoxins: The Taste That Kills*, claims a single meal may contain several excitotoxins and says many parents are unaware of these dangers and unwittingly put their children at risk.

Methanol – When aspartame breaks down in the bloodstream, methanol, a deadly poison, is released. Dr. H.J. Roberts in *Aspartame (NutraSweet) Is it Safe?* claims the American

Environmental Protection Agency recommends an intake of less than eight milligrams of methanol as an acceptable daily level and contends that one litre of diet soda contains approximately fifty-five milligrams. Methanol is said to specifically seek out the blood vessels of the optic nerve and binds itself to it. High levels of methanol destroy the supply of blood to the retina and cause blindness. Signs of methanol poisoning include headaches, numbness of the extremities, dizziness, depression, blurred vision, nausea, vomiting, abdominal pain, seizures, cardiac changes and death.

Phenylalanine – The major component of aspartame is the amino acid, phenylalanine. People who suffer from phenylketonuria (PKU) are unable to metabolize phenylalanine and must avoid aspartame. This problem is so serious for this segment of the population that food manufacturers are now required to include a warning on their labels that the product contains phenylalanine. Unfortunately, large numbers of people are carriers of the PKU gene without knowing it and would be unaware of the risks involved in ingesting products containing aspartame. Dr. Ross Hume Hall points out that because of the seriousness of this condition, "...all babies are routinely screened at birth, PKU children are quickly identified. The problem comes with carriers of one defective gene. They can handle the amount of phenylalanine in a normal diet, but excess phenylalanine from aspartame may overload their bodies." Studies are now suggesting that increases in the levels of phenylalanine ingested during pregnancy (whether the woman is a PKU carrier or not) can adversely effect the growth of the baby's brain.

Supporters of aspartame argue that phenylalanine and aspartic acid are amino acids that occur naturally in milk and other proteins and therefore are safe. What they fail to discuss is that these foods also contain many other amino acids which react with each other as nature intended. When food manufacturers use aspartame in a product, they have pulled two amino acids out of their natural form and flooded the bloodstream with them. These two amino acids, now out of their natural environment, begin to break down into toxic products.

The following summarizes how aspartame has been tested, approved, removed, and then approved again.

- Aspartame was discovered in 1965 by a chemist working for the G.D. Searle company. The company conducted studies to determine aspartame's safety prior to seeking approval from the FDA for use in the marketplace.
- In 1969, a study was carried out by Dr. Harry Waisman, Professor of Pediatrics at the University of Wisconsin, at the request of G.D. Searle. In this study, the doctor fed aspartame mixed with milk to infant monkeys. Unfortunately, Dr. Waisman passed away before his studies were finished but it seems the results were later submitted to the FDA and any negative data he had recorded had disappeared.
- In 1974, aspartame received its first approval for limited use based on test results submitted by G.D. Searle. After approval was granted, Dr. John Olney, a research scientist from the Washington School of Medicine in St. Louis, submitted new findings to the Food and Drug Administration. The studies revealed that mice that had been fed aspartic acid (a major component of aspartame) developed holes in their brains.
- In 1975, an FDA task force found evidence that data submitted by G.D. Searle, from Dr. Waisman's research in 1969, had been falsified. The study revealed development of seizures, brain tumours and a dead monkey. Searle's researchers left out negative data and were accused of removing tumours in test animals without reporting them or examining them to ascertain if they were cancerous. Animals that had died during the study were reported to be alive when the results were submitted to the Food and Drug Administration.
- In 1977, another FDA task force was formed to look into the data previously submitted by G.D. Searle. Once more, it was discovered that data had been altered. A number of test animals were found to have developed uterine tumours that were believed to be a result of diketopiperazine, a breakdown product of aspartame. Due to these findings, the FDA ordered a grand jury investigation of the studies done by G.D. Searle on aspartame.
- In 1980, aspartame came up for approval once more. Dr. Goyan, the FDA commissioner, decided not to approve aspartame for the time being.

- In 1981, G.D. Searle tried once more to have aspartame approved. Within a few months Dr. Arthur Hayes, a new FDA commissioner., approved the use of aspartame in dry foods over the objections of many concerned scientists.
- In 1983, Dr. Hayes approved the use of aspartame in diet soft drinks.
- Since 1984, thousands of complaints of adverse reactions to aspartame (also called NutraSweet™ or Equal™) have been reported to the Food and Drug Administration.

As a note of interest, Monsanto (the parent company of G.D. Searle) has already developed the next generation of sweetener to replace aspartame. It is presently being called neotame and is said to be about eight thousand times sweeter than sugar. Due to its immense sweetening power, it will have to be mixed with bulking agents. Nick Rosa, the president of Monsanto's nutrition and consumer business unit, has said that neotame is based on the aspartame formula but he would not give any further information. Neotame's association with aspartame should automatically make it suspect in the minds of consumers.

Azodicarbonamide – may be used in bread, flour and whole wheat flour as a bleaching, maturing and/or dough-conditioning agent. Although it is considered to have no effect on the nutritional value of vitamins and amino acids in bread, the FDA requires that it be tested for short and long-term effects. In 1958, the use of chemical oxidizing agents, such as azodicarbonamide, was banned in Germany. Authorized for use in Canada and the United States.

Baking soda – *See* Sodium bicarbonate.

Beef tallow – *See* Tallow flakes.

Benzoate of soda – *See* Sodium benzoate.

Benzoic acid – may be used in fruit juices, jams, jellies, marmalades, mincemeat, pickles, relishes, tomato catsup, tomato paste and certain unstandardized foods as a Class II preservative. It is found naturally in cherry bark, cassia bark and to an appreciable degree in most berries. It should be noted that the chemical additive used in food is a carboxylic acid produced from benzene. Benzoic acid can be used as an antifungal agent, chemical preservative and as a dietary supplement up to 0.1 per cent.

Benzoic acid has been known to cause allergic reactions such as red eyes, skin rashes and asthma. It may also cause behavioural changes, intestinal upsets and water retention. People who are sensitive to acetylsalicylic acid (ASA) and tartrazine and those who have liver problems should avoid this additive.

Scientists found that benzoic acid impaired the absorption of nutrients and impeded growth in rats. When given in high doses, it caused brain damage and neurological problems such as irritation, ataxia (muscular coordination failure) and convulsions. Tests done in the former Soviet Union reaffirmed these results and indicated that it may be a co-carcinogen. Their tests also showed that combining the preservatives benzoic acid and sodium bisulphite greatly increased the toxic effects, an indication that additives do have a cumulative effect in the human body. In addition, benzene has been linked to leukemia and certain other blood disorders in humans and animals.

The forty-first report of the FAO/WHO Expert Committee on Food Additives, 1993, noted an absence of studies on the effects of benzyl alcohol, benzoic acid, benzyl acetate, benzaldehyde and the benzoate salts on reproductive systems and developing embryos. They recommended that a full review be carried out in 1995 to determine whether these or other studies are required. Authorized for use in Canada and GRAS in the United States.

Benzoyl peroxide – may be used in flour and whole wheat flour as a bleaching, maturing and/or dough-conditioning agent. Flour can be bleached within twenty-four hours with the application of benzoyl peroxide. Due to limited data available, it appears that further research and study is needed to determine the long-term effects of this agent. In 1958, the use of chemical oxidizing agents, such as benzoyl peroxide, was banned in Germany. Authorized for use in Canada and GRAS in the United States.

Benzyl alcohol – may be used in flavour (naming the flavour) and unstandardized flavouring preparations as a carrier or extraction solvent. Benzyl alcohol is used industrially as a solvent in perfumes, local anesthetic and local antiseptic. Consumption of large doses may cause vomiting, diarrhea and central nervous system depression. It has also been known to cause irritation and corrosion of the skin and mucous membranes. The forty-first report of the FAO/WHO Expert Committee on Food Additives, 1993, noted an absence of studies for benzyl alcohol, benzoic acid, benzyl acetate, benzaldehyde and the benzoate salts on reproductive systems and developing embryos and recommended that a full review be performed in 1995 to determine whether these or other studies are required. Authorized for use in Canada and the United States.

Beta carotene – *See* Carotene.

BHA and BHT (Butylated hydroxyanisole and Butylated hydroxytoluene) – may be used in fats, oils, lard, mono- and diglycerides, margarine, shortening, dried breakfast cereals, dehydrated potato products, chewing gum, essential oils, citrus oils, citrus oil flavours, dry flavours, partially defatted pork and beef fatty tissue and certain unstandardized foods as preservatives. It is also permitted in vitamin A liquids and dry vitamin D preparations for addition to food, which means that it could end up in milk. Since ingredients contained in vitamin preparations do not need to appear on labels, the only way to know what is in your milk is to ask your local dairy if their vitamin preparations contain BHA or BHT.

"BHA or BHT in packaging only" is a misleading phrase as the chemicals can migrate onto the food itself. Since many food producers do not use these additives and still produce a good tasting, nutritious and economical product, it is clear that BHA and BHT are non-essential and do not benefit the consumer. Testing in rats showed that BHA and BHT inhibited growth, caused weight loss, damaged the liver, kidneys and testicles, caused baldness and elevated the blood cholesterol levels. It also caused their offspring to be born without eyes and produced chemical defects in the brain. Researchers suspect that these two related preservatives cause cancer and allergic reactions. In addition, studies have revealed that BHT accumulates in greater concentrations in human fat tissue than it does in the fatty tissue of rats. The sudden loss of body fat due to dieting or illness can result in the release of this substance in the body in toxic amounts.

The *Food and Drugs Act*, Table XI, Part IV, permits the following levels of BHA and BHT:

- in essential oils, citrus oil flavours and dry flavours – up to a maximum level of 0.125 per cent.
- in citrus oils – up to a maximum level of 0.5 per cent.

Experiments at Concordia University showed that infant mice exposed to BHA and BHT during pregnancy had abnormal brains and abnormal behaviour patterns. The behavioural changes included increased exploration, sleeplessness, poor grooming habits, decreased reflexes, aggression and severe learning problems. Dr. Bernard Weiss, of the University of Rochester's School of Medicine, was especially concerned with the behavioural problems cited in numerous other research projects. He felt that governments should test additives not only for physical damage but also for behavioural effects. For example Dr. Weiss pointed out that if thalidomide had lowered children's learning potential rather than causing gross deformities, it never would have been detected.

Some of the more obvious effects of BHA and BHT in humans are blisters on the skin, exhaustion, extreme weakness, tightness in the chest, tingling sensations in the face and hands, swelling of the lips and tongue and difficulty in breathing, especially for people who consume large amounts of BHA and BHT. Consumers should question any chemical that has not been sufficiently studied, that accumulates in body fat and is an unnecessary ingredient in food. They would be wise to buy an alternate product and refrain from purchasing any foods which contain these additives.

The government has been remiss in its duty by failing to demand lifetime studies of BHA and BHT on a variety of species. BHT has been banned for use in food in England, Romania, Sweden and Australia. Additives which are banned in one or more countries due to adverse test results should automatically be banned in Canada. Authorized for use in Canada and GRAS in the United States.

Bicarbonate of soda – *See* Sodium bicarbonate.

Blue #1 (Brilliant blue) – *See* Artificial colour.

Blue #2 (Indigotine) – *See* Artificial colour.

Brilliant blue (Blue #1) – *See* Artificial colour.

Bromelain (Bromelin) – may be used in beer, bread, flour, whole wheat flour, meat cuts, waffles and pancakes as a food enzyme. It is produced commercially from certain varieties of pineapple where it occurs naturally. Bromelain is a protein-digesting and milk-clotting enzyme which can be used in the manufacture of wound-debriding agents. Since many enzymes are now being taken from genetically altered sources, there is no guarantee that bromelain will not be genetically engineered also. Please see Genetically Engineered Foods, page 109. Authorized for use in Canada and the United States.

Bromelin – *See* Bromelain.

Brominated vegetable oil (BVO) – may be used in flavour (naming the flavour) and in some beverages as a density-adjusting agent.

Bromine is a toxic, volatile, corrosive liquid which is added to certain vegetable oils to produce brominated vegetable oil. It is used in some soft drinks and carbonated and non-carbonated fruit-flavoured drinks to keep the flavouring oils evenly distributed and to give the drinks a slightly cloudy appearance which makes them look more like fruit juice.

The results of experiments conducted by the British Industrial Biological Research Association (BIBRA) in 1969 – 70, coupled with tests done by the Canadian Food and Drug Directorate (CFDD), brought damaging evidence against brominated vegetable oil. The BIBRA studies showed that rats stored bromine in their tissues in amounts in proportion to the given dosage. All dosage levels produced fat deposits in the liver and higher levels caused fat to deposit in the kidneys and heart. The spleen, pancreas, lung, aorta and skeletal muscles were also affected. All attempts to eliminate the bromine from the tissues and organs of the test animals were ineffectual and a "no harmful effect level" was impossible to establish.

In the Canadian studies, researchers noted extensive biochemical and pathological effects on rats for eighty days after the rodents were fed bromine. Brominated vegetable oil caused growth retardation, impaired metabolism of nutrients, mild anemia and enlargement of the liver, kidneys, spleen and heart. All the animals displayed signs of thyroid problems, inflammation of the muscular walls of the heart, fat deposits in the liver and retarded development of the testicles.

In 1970, FAO/WHO recommended that it not be used as a food additive.

Brominated vegetable oil was banned in Belgium in 1967, in Sweden in 1968 and in Great Britain in 1970. In spite of this, the Canadian and US governments still allow manufacturers to use this additive.

Butane – *See* n-Butane.

Butter yellow – banned in 1919.

Butylated hydroxyanisole (BHA) – *See* BHA and BHT.

Butylated hydroxytoluene (BHT) – *See* BHA and BHT.

Butylene glycol (1,3-Butylene Glycol) – may be used in flavour (naming the flavour) and unstandardized flavouring preparations as a carrier or extraction solvent. In experiments on animals, it has caused transient stimulation of the central nervous system followed by depression, vomiting, drowsiness, coma, respiratory failure, convulsions, kidney damage and death. It is authorized for use in the United States. In Canada, it is governed by the Good Manufacturing Practice regulation.

BVO – *See* Brominated Vegetable Oil.

Caffeine (Caffeine Citrate) – may be used in cola-type beverages to characterize or typify the product. It is a stimulant, is known to be addictive and occurs naturally in tea leaves, coffee, cocoa and kola nuts. It is interesting to note that even though caffeine is present naturally in the kola nuts, more caffeine is added to cola drinks. One 23dl (eight-ounce) glass of a cola drink can contain up to seventy-two milligrams of caffeine while an equivalent amount of coffee contains approximately ninety milligrams of caffeine.

Most people are aware of some of caffeine's common adverse effects such as sleeplessness, irritability, ringing in the ears, heart palpitations and ulcer aggravation. However, most people don't realize that consuming large amounts of caffeine over long periods of time can have serious toxic effects, such as elevated blood pressure and heart problems, central nervous system damage, behavioural changes, changes in blood sugar level, reduced fertility, miscarriages, fibrocystic breasts, anxiety, panic attacks and birth defects. Large quantities of caffeine can seriously effect children as their small bodies are much more sensitive to toxic substances. It is known to cause hyperactivity and nervousness in children as well as in adults.

In test-tube experiments involving cells taken from mice and humans, DNA repair was suppressed, proving that caffeine has a mutagenic effect. Caffeine is capable of reaching the forming fetus; birth defects could occur with as little as one or two cups of coffee a day. Laboratory testing of animals indicates a real danger to developing fetuses, especially in the first three months when the vital organs are being formed. Doctors should inform expectant mothers of this fact. As a precaution, caffeine intake should be reduced or totally eliminated for the duration of the pregnancy. This applies to decaffeinated coffee also – it can be even more dangerous, since powerful chemicals are used to extract the caffeine. Residues of these chemicals can be deposited in the brain, liver and muscles. *See* Dichloroethane (Ethylene dichloride) and Dichloromethane (Methylene chloride).

Look for decaffeinating methods that don't use chemical solvents. One method applies liquid carbon dioxide at high pressure to extract the caffeine from the beans and leaves no harmful residues. Another method, known as the Swiss Water Process, uses charcoal filters to remove caffeine from water in which the beans have been soaked. Any decaffeinated coffee can be referred to as "water-processed," but if you wish to avoid solvents, you need to look particularly for "Swiss Water Processed." Authorized for use in Canada and the United States.

Caffeine citrate – *See* Caffeine.

Calcium acetate – may be used in ale, beer, light beer, malt liquor, porter, stout and unstandardized foods as a pH adjusting agent, acid-reacting material and/or water-correcting agent. Industrially, it is employed in dyeing, tanning and curing skins, in lubricants and as a corrosion inhibitor. It is said to have low toxicity when taken orally. Authorized for use in Canada and GRAS in the United States.

Calcium alginate – *See* Alginate.

Calcium aluminum silicate (Aluminum calcium silicate) – may be used in salt, garlic salt and unstandardized dry mixes as an anti-caking agent. One of its industrial uses is as a component of cement. Authorized for use in Canada and GRAS in the United States. *See* Silicate.

Calcium ascorbate (Ascorbic acid calcium salt) – may be used in ale, beer, wine, canned mushrooms, canned tuna, cider, frozen fruit, preserved fish, preserved meat or meat by-products, canned apple-sauce, canned peaches, fats, oils, lard, mono- and diglycerides, diglycerides, shortening and unstandardized foods as a preservative. This chemical is prepared from ascorbic acid and calcium carbonate. Authorized for use in Canada and GRAS in the United States.

Calcium carbonate – may be used in bread and unstandarized bakery foods as a yeast food; in flour and whole wheat flour as a carrier of benzoyl peroxide; in confectionery as a creaming and fixing agent; in chewing gum as a filler; in unstandardized foods as a carrier and dusting agent; in ice cream mix, cream cheese spreads, grape juice and unstandardized foods as a pH adjusting agent, acid-reacting material and/or water-correcting agent.

It is found naturally in limestone, marble and coral and is more commonly known as chalk. Calcium carbonate is used to manufacture various industrial products such as paint, rubber, plastics and paper. It is also used as a gastric antacid and anti-diarrheal medicine. Authorized for use in Canada and GRAS in the United States.

Calcium carrageenan – *See* Carrageenan.

Calcium chloride – may be used in tomatoes, canned vegetables, frozen and canned apples, canned grapefruit, cheese, cottage cheese, olives, pickles, relishes and unstandardized foods as a firming agent; in ale, beer, porter, malt liquor, stout and unstandardized foods as a pH adjusting agent, acid-reacting material and/or water-correcting agent and in unstandardized bakery foods as a yeast food. It is employed industrially in the manufacture of antifreeze, refrigerating solutions, ice, glue, cements, in fire extinguishers and in wood and stone preservation. Medically, calcium chloride is used to replace the body's electrolytes, to fight allergies and to make urine more acidic. It has been known to cause irregular heart beats and upset stomachs. Authorized for use in Canada and GRAS in the United States.

Calcium citrate – may be used in infant formula and unstandardized foods as a pH adjusting agent, acid-reacting material and/or water-correcting agent; in processed cheese, cream cheese spread and unstandardized foods as an emulsifying, gelling, stabilizing and/or thickening agent; in tomatoes, canned vegetables, frozen apples, sliced apples, canned apples and unstandardized foods as a firming agent; in unstandardized foods as a sequestering agent and in unstandardized bakery foods as a yeast food. Citrates have been known to interfere with the results of certain laboratory tests, such as tests for abnormal liver function, pancreatic function and blood pH levels. Authorized for use in Canada and GRAS in the United States.

Calcium cyclamate – *See* Cyclamate.

Calcium dioxide – *See* Calcium peroxide.

Calcium disodium EDTA (Disodium ethylenediaminetetraacetate, Disodium EDTA) – may be used in ale, beer, light beer, malt liquor, porter, stout, salad dressing, mayonnaise, unstandardized dressings and sauces, potato salad, sandwich spread, (canned) shrimp, tuna, clams, lobster and salmon, margarine, soft drinks and ready-to-drink teas as a sequestering agent. Medically, it is used as a detoxicant to treat poisonings by lead or other heavy metals. Its use in this capacity in humans has produced clear evidence of kidney damage.

Food manufacturers use calcium disodium EDTA to bind with metals which may enter food from mixing containers. These metals could change the texture, colour or flavour of the foods. In excessive amounts, calcium disodium EDTA can cause a mineral imbalance in the body as it traps calcium, iron and other nutrients and prevents the body from using them. It is known to inhibit blood-clotting and can cause muscle cramps, intestinal upsets and blood in the urine. In animal experiments, it caused disruption of the cellular metabolism leading to chromosome damage. This additive is particularly dangerous because it is ingested by consumers in so many foods and beverages every day. Calcium disodium EDTA is on the FDA's priority list to be studied for mutagenic, teratogenic, subacute and reproductive effects. Authorized for use in Canada and the United States.

Calcium furcelleran – *See* Furcelleran.

Calcium gluconate – may be used in unstandardized foods as an emulsifying, gelling, stabilizing, thickening, firming, pH adjusting, acid-reacting material and/or water-correcting agent. It is employed industrially to purify sewage. Medically, it is used as a calcium replenisher. It was decided by the FDA in 1980 to continue its GRAS status under the Good Manufacturing Practice regulation. Authorized for use in Canada.

Calcium glycerophosphate – may be used in unstandardized dessert mixes as an emulsifying, gelling, stabilizing and/or thickening agent. It can be employed commercially to manufacture dentrifices and is used in dietary supplements. Medically, it is administered to alleviate numbness and fatigue. In 1980, the FDA renewed its GRAS status. Authorized for use in Canada. *See also* Phosphoric acid.

Calcium hydrate – *See* Calcium hydroxide.

Calcium hydroxide (Calcium hydrate) – may be used in infant formula, ale, beer, light beer, malt liquor, porter, stout, ice cream mix, canned peas, grape juice and unstandardized foods as a pH adjusting agent, acid-reacting material and/or water-correcting agent. It is an alkali commonly known as slaked lime and is commercially prepared by hydration of lime. It is used industrially in mortar, plaster, cement and other building materials. Calcium hydroxide is also used as a topical astringent for medical purposes. At present, it is considered non-toxic although it is a potent caustic in its undiluted form. Authorized for use in Canada and GRAS in the United States.

Calcium hypophosphite – may be used in unstandardized dessert mixes as an emulsifying, gelling, stabilizing and/or thickening agent. It is used industrially as a corrosion inhibitor and in nickel plating. In 1980, the FDA renewed its GRAS status. Authorized for use in Canada.

Calcium iodate – may be used in bread and unstandardized bakery foods as a bleaching, maturing and/or dough-conditioning agent. It is used in animal feed as an iodine supplement. Medically, it is used as a topical disinfectant. Authorized for use in Canada and the United States.

Calcium lactate – may be used in canned grapefruit and canned peas as a firming agent; in baking powder and unstandardized foods as a pH adjusting, acid-reacting material and/or water-correcting agent; in bread and unstandardized bakery foods as a yeast food. Medically, it is used as a calcium replenisher and commercially it is employed to manufacture dentrifices. Authorized for use in Canada and GRAS in the United States.

Calcium oxide – may be used in ale, beer, light beer, malt liquor, porter, stout, ice cream mix and unstandardized foods as a pH adjusting agent, acid-reacting material and/or water-correcting agent. Because it is commercially produced from limestone, it is commonly known as lime. It is used industrially in bricks, dehairing hides, insecticides and various other products. Direct contact with calcium oxide can cause severe skin and mucous membrane damage as it is a potent caustic when dissolved in water, where it forms calcium hydroxide. Authorized for use in Canada and GRAS in the United States.

Calcium peroxide (Calcium dioxide) – may be used in bread and unstandardized bakery foods as a bleaching, maturing and/or dough-conditioning agent. Industrially, it is also used as a stabilizer for rubber. Calcium peroxide is employed medically as an antiseptic. Authorized for use in Canada and the United States. *See also* Hydrogen peroxide.

Calcium phosphate, dibasic (Dicalcium phosphate) – may be used in processed cheese, processed cream cheese, skim milk processed cheese and unstandardized foods as an emulsifying, gelling, stabilizing and/or thickening agent; in unstandardized foods as a firming agent, pH adjusting agent, acid-reacting material and/or water-correcting agent; and/or in flour and whole wheat flour as a carrier of benzoyl peroxide and of potassium bromate; and in bread and unstandardized bakery foods as a yeast food. It is also employed in animal feeds, dental products, fertilizers and the production of glass. Medically it is used as a mineral supplement. At present, this additive is considered non-toxic. Authorized for use in Canada and GRAS in the United States. *See also* Phosphoric acid.

Calcium phosphate, monobasic (Monocalcium phosphate) – may be used in tomatoes, canned vegetables, frozen and canned apples and unstandardized foods as a firming agent; in ale, beer, light beer, malt liquor, porter, stout, baking powder and unstandardized foods as a pH adjusting agent, acid-reacting material and/or water-correcting agent; in ice cream mix, sherbet and unstandardized dairy products as a sequestering agent and in bread, flour and unstandardized bakery foods as a yeast food. It is also utilized in fertilizers and is a mineral supplement in foods and feeds. At present, this additive is considered non-toxic. Authorized for use in Canada and GRAS in the United States. *See also* Phosphoric acid.

Calcium phosphate, tribasic (Tricalcium phosphate) – may be used in salt, garlic salt, onion salt, dry cure, oil-soluble annatto, icing sugar and unstandardized dry mixes as an anti-caking agent; in unstandardized foods as an emulsifying, gelling, stabilizing and/or thickening agent and/or as a pH adjusting agent, acid-reacting material and/or water-correcting agent; in flour and whole wheat flour as a carrier of benzoyl peroxide; in ice cream mix and ice milk mix as a sequestering agent and in unstandardized bakery foods as a yeast food. Industrially it is employed in the manufacture of fertilizers, milk-glass, polishing and dental powders, porcelains, pottery and animal feeds. Calcium phosphate tribasic is used medically as a gastric antacid and mineral supplement. At present, this additive is considered non-toxic. Authorized for use in Canada and GRAS in the United States. *See also* Phosphoric acid.

Calcium phytate – may be used in glazed fruit as a sequestering agent. The FDA's Bureau of Medicine went on record as saying "...we do not actually recognize calcium phytate as necessarily safe..." In 1980 the FDA decided to continue its GRAS status under the Good Manufacturing Practice regulation. Authorized for use in Canada.

Calcium propionate – may be used in bread, cheese and certain unstandardized foods as a Class III preservative. Medically it is used as an antifungal agent. Commercially it is used as a preservative to inhibit the growth of molds and other micro-organisms in tobacco and pharmaceuticals. Authorized for use in Canada and GRAS in the United States. *See also* Propionic acid.

Calcium silicate – may be used in salt, garlic salt, onion salt, baking powder, dry cure, grated cheese, dried whole egg, dried egg yolk, dried whole egg mix, unstandardized dry mixes and icing sugar as an anti-caking agent; and in oil-soluble annatto as a carrier. Industrially, it is employed in road construction. Its toxicity when taken orally is suggested to be virtually nil although it has been known to irritate the respiratory tract when it is inhaled in a concentrated form. Listed by the FDA for further study of its mutagenic, teratogenic, subacute and reproductive effects. Authorized for use in Canada and GRAS in the United States. *See also* Silicate.

Calcium sorbate – may be used in apple or rhubarb jam, jellies with pectin, mincemeat, pickles, relishes, smoked or salted dried fish, tomato catsup, tomato paste, tomato purée, olive brine, margarine, unstandardized salad dressings and unstandardized foods as a Class II preservative, and is used in bread, cheese, cheddar cheese, cream cheese, cider, wine and honey wine as a Class III preservative. Authorized for use in Canada and GRAS in the United States. *See also* Sorbate.

Calcium stearate – may be used in free-running salt, flour salt, garlic salt, onion salt and unstandardized dry mixes as an anti-caking agent; and in confectionery as a release agent. It is employed industrially for waterproofing fabrics, cement, stucco and explosives, and is used in the manufacture of pencils and wax crayons. Authorized for use in Canada and GRAS in the United States.

Calcium stearoyl-2-lactylate – may be used in bread, unstandardized bakery foods and cake mixes as a bleaching, maturing and/or dough-conditioning agent; in liquid and frozen egg whites, dried egg whites and vegetable fat toppings as a whipping agent and in dehydrated potatoes as a conditioning agent. Although some scientific tests suggest this additive is harmless, no lifetime feeding studies or reproduction studies have been conducted. Presently, it is on the FDA's list requiring further information. Authorized for use in Canada and the United States.

Calcium sulphate – may be used in ice cream, ice milk mix, sherbet, creamed cottage cheese and unstandardized foods as an emulsifying, gelling, stabilizing and/or thickening agent; in tomatoes, canned vegetables and frozen and canned apples as a firming agent; in flour and whole wheat flour as a carrier of benzoyl peroxide; in baking powder as a neutral filler; in ale, beer, light beer, stout, malt liquor, porter, wine as a pH adjusting agent, acid-reacting material and/or water-correcting agent and in bread and unstandardized foods as a yeast food. Industrially it is employed in wall plaster, cement and insecticides. This chemical, commonly known as plaster of Paris, absorbs water and hardens quickly, making it ideal for its medical use in plaster casts. Unfortunately, also due to these properties, its ingestion can cause obstruction of the bowels. As a point of interest, a mixture of calcium sulphate and flour is used to kill rodents. Authorized for use in Canada and GRAS in the United States.

Calcium tartrate – may be used in or upon unstandardized foods as an emulsifying, gelling, stabilizing and/or thickening agent. It is employed medically as an antacid. Calcium tartrate is a wine industry by-product and is prepared commercially from wine dregs. To date, it is not considered toxic. Authorized for use in Canada and the United States.

Cane syrup – *See* Sugar.

Canthaxanthin – *See* Artificial colour.

Caramel – may be used in ale, beer, light beer, porter, stout, jams, jellies, brandy, bread, brown bread, butter, cider vinegar, ice cream mix, icing sugar, liqueurs, milk (naming the flavour), mincemeat, pickles and relishes, rum, sherbet, smoked fish, lobster paste, tomato catsup, whiskey, wine, honey wine, caviar and unstandardized foods as a colouring agent.

When we wrote to one food company inquiring about caramel, we were quickly reassured that "caramel colouring is produced by caramelization of liquid corn syrup. It is a totally natural ingredient – no preservatives, stabilizers or additives are used. It produces flavour as well as colour. Most dark breads contain it because the consumers demand that colour." We later discovered that the colour additive "caramel" is the dark-brown liquid or solid material resulting from the carefully controlled heat treatment of the following food-grade carbohydrates: dextrose, invert sugar, lactose, malt syrup, molasses, starch hydrolysates and sucrose. Approximately nineteen chemicals can be used to assist caramelization, of which the following are a few: potassium hydroxide, sodium hydroxide, ammonium hydroxide, calcium hydroxide and ammonia compounds.

In 1969, the WHO said the use of ammonia compounds in the production of caramel for food was unacceptable. Their findings showed the effects of a nerve toxin found in caramel produced convulsions in test animals. Canada reduced the limits of ammonia compounds manufacturers could use in caramel production to a level considered acceptable by the World Health Organization.

However, a child of sixteen kilograms (35 lbs) who drank two cans of soft drinks a day that contained caramel, would exceed the acceptable daily limit. This doesn't include the amount of caramel a child may get each day in the previously mentioned foods. This negates the government's policy regarding "parts per million" which allows manufacturers to continue using toxic chemicals in food under the guise of a "no harmful effect" limit. Many prominent scientists reject the "parts per million" standard as they feel there is no acceptable level for a dangerous chemical.

Research indicated that caramel, in which ammonia compounds had been used, reduced the number of white blood cells in rats. It also caused a decrease in the weight of the spleen and thymus, which are connected with disease immunity. An adult would exceed the acceptable level by drinking two 355-millilitre (12-ounce) bottles of stout, porter or bock beer daily.

Caramel is on the FDA's list for further study of mutagenic, teratogenic, subacute and reproductive effects. The *Food and Drugs Act* regulates caramel under the Good Manufacturing Practice regulation, which permits the food industry to add this chemical in any amount they require to achieve a desired result. Therefore, even though the level of ammonia compound in caramel has been reduced, manufacturers are still permitted to add as much caramel as they wish to the aforementioned products. The ammonia compound levels would differ substantially in different products. Authorized for use in Canada and GRAS in the United States.

Carbon black – *See* Artificial colour.

Carbon dioxide – may be used in ale, beer, light beer, malt liquor, porter, stout, wines, carbonated juice, cider, mineral water and in unstandardized foods as a carbonation and pressure-dispensing agent. This additive is governed by the Good Manufacturing Practice regulation which gives industry control over the amount of carbon dioxide it uses in products. It is an odourless gas used as dry ice for refrigeration and, on stage, produces harmless smoke. Carbon dioxide has been known to cause shortness of breath, vomiting, elevated blood pressure and confusion. Authorized for use in Canada and GRAS in the United States.

Carboxymethylcellulose – may be used in cream, French dressing, milk, mustard pickles, relishes, salad dressing, cottage cheese, ice cream, ice milk, sherbet, processed cheeses, cream cheese and cream cheese spreads as an emulsifying, gelling, stabilizing and/or thickening agent and in sausage casings to enable peeling. It is a white powder produced from cotton by-products. This additive, when ingested, has been found to cause cancer in animals. Authorized for use in Canada and GRAS in the United States. *See also* Sodium carboxymethylcellulose.

Carmine – *See* Cochineal.

Carnauba wax – may be used in confectionery as a glazing and polishing agent. It is obtained from the leaves and buds of a South American palm tree. Other uses of carnauba wax include polishes, varnishes, lubricants and cosmetic products. The FDA has listed it for further testing of its mutagenic, teratogenic, subacute and reproductive effects. To date the amount of testing done on this additive is insufficient to evaluate its toxicity with any certainty. Authorized for use in Canada and the United States.

Carob bean gum (Locust bean gum) – may be used in cream, French dressing, salad dressing, milk, mustard pickles, processed cheese, cottage cheese, ice cream, ice milk, calorie-reduced margarine, sherbet, sour cream, cream cheese, relishes and unstandardized foods as an emulsifying, gelling, stabilizing and/or thickening agent. It is obtained from the Mediterranean carob tree. Studies suggest that it may lower blood cholesterol levels. It is now listed by the FDA for testing of its mutagenic, teratogenic, subacute and reproductive effects. Authorized for use in Canada and GRAS in the United States.

Carotene (Beta carotene) – is permitted in jam, bread, butter, cheese, concentrated fruit juice except frozen concentrated orange juice, liquid, dry or frozen whole egg, marmalade, ice cream mix, icing sugar, margarine, jellies, liqueurs, pickles, relishes, sherbet, milk, smoked fish, lobster paste, caviar, tomato catsup and unstandardized foods as a colouring agent. Carotene is a yellow-to-orange pigment derived from plant sources. Beta carotene, or previtamin A, is one of several forms of carotene. Beta carotene is converted to vitamin A in the liver. It is found in many green, yellow and orange fruits and vegetables and scientists say it is completely non-toxic. Too much beta carotene will simply cause a yellowish discolouration of the skin which quickly subsides by reducing its intake.

Beta carotene is unnecessary in foods as a colouring agent and vitamin A supplements should not be required (except on doctor's orders) as most people get more than enough in even the simplest diet of unprocessed foods. Authorized for use in Canada and GRAS in the United States.

Carrageenan (Ammonium carrageenan, Calcium carrageenan, Potassium carrageenan, Sodium carrageenan) – may be used in infant formula, cream, French dressing, head cheese, jelly with pectin, cottage cheese, ice cream, ice milk, evaporated milk, sherbet, canned poultry, calorie-reduced margarine, sour cream, canned asparagus, waxed beans, peas, green beans, meat loaf, mustard pickles, relishes, salad dressing, porter, beer, malt liquor, ale, light beer and unstandardized foods as an emulsifying, gelling, stabilizing and/or thickening agent. It is obtained from red seaweed commonly found in the North Atlantic coastal regions from Norway to North America. Due to its plant origin and its long-time use, carrageenan was considered safe. However, in 1969, experiments showed that it caused tiny ulcers in the large intestine, similar to those found in humans with ulcerative colitis. Other adverse effects were blood and mucus in the feces and stunted growth. Scientists feel that the public is possibly being subjected to a danger of unknown dimension. Due to the fact that carrageenan has no nutritional value, and studies have shown it may pose serious health hazards, it should not be used in food, especially infant formulas. Infant formula in Britain does not include carrageenan. Authorized for use in Canada and GRAS in the United States.

Castor oil – may be used in confectionery as a releasing or anti-sticking agent and in annatto butter and margarine colour as a carrier or extraction solvent. It is also used in embalming fluid and medically as a cathartic and topical emollient. It is obtained from castor oil plant seeds. Castor oil, in large quantities, has been known to cause pelvic congestion and induce abortions in humans. Authorized for use in Canada and the United States.

Catalase – may be used in soft drinks and egg albumen as a food enzyme and is taken from bovine liver or the fungus *Aspergillus niger*. In combination with glucose oxidase, it is used to treat food wrappings to prevent food deterioration. Since many enzymes are now being taken from genetically altered sources, there is no guarantee that catalase will not also be genetically engineered. For more information on genetic engineering, please see Genetically Engineered Foods, page 109. Authorized for use in Canada and the United States.

Cellulose gum – may be used in cream, French dressing, milk, mustard pickles, relishes, salad dressing, cottage cheese, ice cream, ice milk, sherbet, processed cheeses, cream cheese and cream cheese spreads as an emulsifying, gelling, stabilizing and/or thickening agent and in sausage casings to enable peeling. It is a white powder produced from cotton by-products. Industrially, it can be used in varnishes, lacquers, coatings on airplane fabrics and in the manufacture of rubber. In 1980, the FDA renewed its GRAS status for packaging only. Authorized for use in Canada. *See* Sodium carboxymethylcellulose.

Cellulose microcrystalline – *See* Microcrystalline cellulose.

Charcoal – *See* Artificial colour.

Chlorine (Chlorine gas) – may be used in flour and whole wheat flour as a bleaching, maturing and/or dough-conditioning agent at levels consistent with the Good Manufacturing Practice regulation. Rather than naturally aging flour by storing it for several months, manufacturers find it more economical to speed up the process by using an oxidizing agent such as chlorine. It is blasted on the flour in the final stage of production. Chlorine has a suffocating odour and is a powerful irritant which is dangerous and possibly lethal to inhale. Carbon tetrachloride, a carcinogen, is formed when chlorine reacts with organic substances. Chlorine has been used as a poison gas under the name Bertholite. In 1958, the use of chemical oxidizing agents, such as chlorine, was banned in Germany. Authorized for use in Canada and the United States.

Chlorine dioxide – may be used in flour and whole wheat flour as a bleaching, maturing and/or dough-conditioning agent. A major part of the vitamin E occurring naturally in these foods is destroyed by the use of chlorine dioxide. It is capable of causing erosion of the skin and mucous membranes of the respiratory tract which could lead to pulmonary edema. In 1958, the use of chemical oxidizing agents, such as chlorine dioxide, was banned in Germany. Authorized for use in Canada and the United States.

Chlorophyll – permitted for use in bread, butter, concentrated fruit juice (unfrozen varieties), ice cream mix, icing sugar, jam with pectin, liqueurs, milk, pickles, relishes, sherbet, skim milk, smoked fish, lobster paste, caviar, tomato catsup, liquid, dried or frozen whole egg or egg yolk, vegetable fats and oils, margarine, a variety of cheeses and unstandardized foods as a colouring agent. It is the green colouring material in plants which plays a vital role in the process of photosynthesis. At present, it is considered non-toxic. Authorized for use in Canada and the United States.

Citric acid – may be used in jams, jellies, salad dressing, infant formula, margarine, canned applesauce, grapefruit, pears and pineapple as a pH adjusting agent, acid-reacting material and/or water-correcting agent and in fats, oils, lard, shortening and some unstandardized foods as a preservative. It occurs naturally in high concentrations in citrus fruits and berries. However, it is produced synthetically by using specific moulds that, under certain conditions, will change sugar to citric acid. Manufacturers then are able to extract the citric acid in its pure form. Research to date indicates no known toxicity other than some allergic responses. Authorized for use in Canada and GRAS in the United States.

Citrus red #2 – *See* Artificial colour.

Coal tar dye – *See* Artificial colour.

Cochineal (Carmine) – may be used in bread, butter, concentrated fruit juice (unfrozen varieties), ice cream mix, icing sugar, jam with pectin, liqueurs, milk, pickles, relishes, sherbet, skim milk, smoked fish, lobster paste, caviar, tomato catsup, liquid, dried or frozen whole egg, vegetable fats and oils, margarine, a variety of cheeses, unstandardized foods and many other processed foods as a natural colouring agent. This crimson dye is derived from a species of Mexican insect. In the US it was discovered to have caused a severe outbreak of an intestinal infection (salmonellosis) which killed one infant and rendered over twenty other patients seriously ill. The cause of the infection was a cochineal solution which was used by doctors to test patients' digestive organs. Research on long-term effects on reproductive systems and human metabolism is unavailable. Authorized for use in Canada and the United States.

Coconut oil – is not considered an additive according to Division 1 of the *Food and Drugs Act*. It is a highly saturated fat obtained from kernels of the coconut palm and is used in the manufacture of soap, chocolates, self-basting turkeys and candles. Although considered non-toxic, it is thought to contribute to hardening of the arteries. Authorized for use in Canada and GRAS in the United States.

Colour – *See* Artificial colours.

Corn dextrin (Dextri-maltose) – is not considered an additive according to Division 1 of the *Food and Drugs Act*. It may be used as a thickening agent or modifier in milk products. At present, it is considered to be non-toxic. Authorized for use in Canada and GRAS in the United States.

Corn sugar – *See* Sugar.

Corn syrup – *See* Sugar.

Cream of tartar – *See* Sodium potassium tartrate.

Cyclamate (Sodium cyclamate, Calcium cyclamate, Potassium cyclamate) – is a substance thirty times as sweet as sugar and ten times cheaper. Although it is used mainly by people attempting to lose or control weight, a joint study conducted in 1956 by the Harvard School of Public Health and the Peter Bent Brigham Hospital concluded that no significant difference in weight loss resulted between users and non-users of products containing cyclamate.

Cyclamate was first used as a low-calorie sweetener in 1953. As early as 1954, the National Academy of Sciences was warning that cyclamate should not be used by the general public. These warnings were repeated in 1955, 1962 and 1968 but were disregarded by industry, government and consumers. Finally, in 1969, results of tests could no longer be ignored. Cyclamate was found to cause damage to the liver and intestinal tract, bladder cancer, birth defects, mutations and testicular damage in animals. Other studies indicated it was capable of crossing the placental barrier and of being distributed through breast milk. Cyclamate was also noted to affect the absorption of certain drugs. The FDA initially responded by recommending that consumers voluntarily restrict their use of cyclamate. Then as public pressure increased, they decided to ban it in October, 1969. Pressure from industry forced the government to change their position in February 1970 to allow cyclamate to be used as long as it was labelled as a drug. Later that year, public and scientific pressure forced the FDA to capitulate and ban it once again. In 1980, the FDA upheld the ban on cyclamate as the National Research Council had reported that cyclamate affected growth, reproduction and blood pressure in rats. It also appeared to enhance the potency of other carcinogens.

The following excerpts from the *Food and Drugs Act* describe how it is still regulated in Canada.

> E.01.004. (1) Every cyclamate sweetener that is not also a saccharin sweetener shall be labelled to state that such sweetener should be used only on the advice of a physician.

E.01.005. Commencing June 1, 1979, every cyclamate sweetener or saccharin sweetener shall be labelled to show

(a) a list of all the ingredients and, in the case of

(i) cyclohexyl sulphamic acid,

(ii) a salt of cyclohexyl sulphamic acid,

(iii) a saccharin,

(iv) a saccharin salt, or

(v) carbohydrates,

the quantity thereof contained in the sweetener

Unbelievably, cyclamate is still allowed for use in Canada in table-top sweeteners such as Sugar Twin™.

Cysteine – may be used in bread, flour, whole wheat flour and unstandardized bakery foods as a bleaching, maturing and/or dough-conditioning agent. Naturally occurring cysteine is an essential amino acid which is a natural component of foods containing protein. Cysteine is medically employed to aid wound healing. It is on the FDA's additive list to be tested further. Authorized for use in Canada and GRAS in the United States.

Dextri-maltose – *See* Corn dextrin.

Dextrin – is not considered an additive according to Division 1 of the *Food and Drugs Act*. It is a starch that has been treated with acids, alkalis or enzymes. At present, it is considered to be non-toxic other than being linked to the occasional allergic reaction. Authorized for use in Canada and GRAS in the United States.

Dextrose – *See* Sugar.

Diammonium phosphate – *See* Ammonium phosphate, dibasic.

Dicalcium phosphate – *See* Calcium phosphate, dibasic.

Dichloroethane (Ethylene dichloride) – may be used in spice extracts, natural extractives, green coffee beans and tea leaves for decaffeination purposes as a carrier or extraction solvent. Testing has found this additive damaged the kidneys, liver and central nervous system. It is a known carcinogen and it has no nutritional benefit.

According to the thirty-ninth report of the FAO/WHO Expert Committee on Food Additives, 1992, long-term studies in mice and rats demonstrated dichloroethane to be carcinogenic. In female mice, it caused mammary and uterine adenocarcinomas and the compound was found to cross the placental barrier of rats.

Hepatocellular carcinomas were induced in male mice. Its use as an extraction solvent was discouraged by the committee, and it recommended an overall review of chlorinated hydrocarbon solvents used in food processing. Authorized for use in Canada and the United States.

Dichloromethane (Methylene chloride) – may be used in spice extracts, natural extractives, green coffee beans and tea leaves for decaffeination purposes as a carrier or extraction solvent. Testing has found this additive to cause damage to the kidneys, liver and the central nervous system. It is a known carcinogen and it has no nutritional benefit.

According to the thirty-fifth report of the FAO/WHO Expert Committee on Food Additives, the use of dichloromethane as an extraction solvent should be discouraged because of toxicological concerns. The committee recommended an overall review of chlorinated hydrocarbon solvents used in food processing. Authorized for use in Canada and the United States.

Diglyceride – *See* Glyceride.

Dimethylpolysiloxane (Silicone, Methyl polysilicone, Methyl silicone) – is permitted in or upon jams, jellies, marmalade, reconstituted lemon/lime juice, shortening, skim milk powder, fats and oils, pineapple juice, wort used in the manufacture of ale, beer, light beer, malt liquor, porter and stout and unstandardized foods as an anti-foaming agent. Silicone has innumerable uses, including the well-known procedure by which the female breasts are enlarged. Although early tests in humans and animals showed no apparent adverse effects, the public is now well aware of problems associated with the use of silicone in the body. Authorized for use in Canada and the United States.

Dioctylsodium sulpho-succinate (DSS) – may be used in fumaric acid (acidulated dry beverage bases) as a wetting agent and in sausage casings to reduce casing breakage. Studies in 1970 on rats using large amounts of this additive had no adverse effects. At the twenty-second meeting of the FAO/WHO Expert Committee on Food Additives, the committee requested further information on the effects on new-born animals (especially those exposed to DSS through lactation), more long-term studies in a rodent species and additional testing on its pulmonary effects. Authorized for use in Canada and the United States.

Dipotassium phosphate – *See* Potassium phosphate, dibasic.

Disodium EDTA – *See* Calcium disodium EDTA.

Disodium ethylenediaminetetraacetate – *See* Calcium disodium EDTA.

Disodium guanylate – is not considered an additive according to Division 1 of the *Food and Drugs Act*. Although virtually tasteless by itself, it intensifies natural flavours of foods and manufacturers often use it to replace more costly natural ingredients. This additive is considered safe as scientists say they understand the body's utilization of it, but it should be avoided by people suffering from gout and other conditions that are aggrevated by purines. Authorized for use in Canada and in the United States.

Disodium inosinate – is not considered an additive according to the *Food and Drugs Act*. It intensifies natural flavours of foods although it is virtually tasteless by itself and manufacturers often use it to replace more costly natural ingredients. This additive is considered safe as scientists say they understand the body's utilization of it, but it should be avoided by people suffering from gout and other conditions that are aggrevated by purines. Authorized for use in Canada and the United States.

Disodium phosphate – *See* Sodium phosphate, dibasic.

DSS – *See* Dioctylsodium sulpho-succinate.

EDTA – *See* Calcium disodium EDTA.

Epichlorohydrin – may be used in starch as a modifying agent. It is employed industrially as a solvent for natural and synthetic resins, gums, cellulose esters and ethers, paints, varnishes, nail enamels and lacquers and cement for celluloid. In one series of tests, rats administered one milligram per kilogram of body weight died within four days. These results raised the question of cumulative potential in humans. In a two-year study, workers exposed for six months or more to epichlorohydrin were found to have an increased cancer rate. Chronic exposure in humans has been shown to cause kidney damage. Germany regulates this additive as a known carcinogen. Permitted for use in starch in the United States. Authorized for use in Canada under the Good Manufacturing Practice regulation. *See also* Starch.

Erythorbate – *See* Sodium erythorbate.

Erythorbic acid (Isoascorbic acid) – may be used in cider, frozen fruit, head cheese, meat binder for preserved meat and preserved meat by-products, ale, beer, light beer, malt liquor, porter, stout, wine, preserved fish, frozen minced fish, preserved meat/poultry, pumping pickle, cover pickle and dry cure employed in the curing of preserved meat or preserved meat by-product, canned applesauce and unstandardized foods as a preservative. It is considered to be non-toxic. Authorized for use in Canada and GRAS in the United States.

Erythrosine (Red #3) – *See* Artificial colour.

Ethanol – *See* Ethyl alcohol.

Ethyl acetate – may be used in spice extracts, natural extractives, flavours, unstandardized flavouring preparations and in green coffee beans and tea leaves for decaffeination purposes as a carrier or extraction solvent. According to Doctors Cheraskin and Ringsdorf, co-authors of *Psycho Dietetics: Foods as the Key to Emotional Health*, synthetically produced ethyl acetate is most frequently used to clean leather and textiles. These doctors charge that the vapours of this chemical damage the lungs, liver and heart. It is also a known central nervous system depressant. Authorized for use in Canada and GRAS in the United States.

Ethyl alcohol (Ethanol) – may be used in spice extracts, natural extractives, flavours, unstandardized flavouring preparations, colour mixtures and preparations, meat and egg marking inks and in food additive preparations as a carrier or extraction solvent. Two of the methods by which ethyl alcohol can be produced are by fermentation of starches and sugars and by processing from ethylene, acetylene and sulphite waste liquors. Manufacturers can aid fermentation by using various chemicals. Ethyl alcohol is the alcohol commonly found in beer, wine and liquor. It is rapidly absorbed through the gastrointestinal tract and ingestion of large doses over a short period of time (i.e., minutes) causes alcohol poisoning resulting in nausea, vomiting, dizziness, loss of balance, convulsions, coma and even death. In 1976, ethyl alcohol was also approved for use in the US in pizza crust to extend handling and storage life. Authorized for use in Canada and the United States.

Ethylene dichloride – *See* Dichloroethane.

Ethylene gas – *See* Ethylene oxide.

Ethylene oxide (Ethylene gas) – may be used as a fumigant for whole or ground spices, except mixtures containing salt. Synthetic ethylene gas is also employed commercially to accelerate the ripening of fruit. For instance, bananas can be picked green and gassed so they will be a nice yellow on the outside. However, the fruit remains immature and hard and lacks the maximum nutritional quality of sun-ripened fruit. In addition, ethylene oxide destroys certain B vitamins and amino acids. In studies, rats fed gas-treated diets were found to have arrested growth. Perhaps a more serious implication is the fact that ethylene oxide reacts with the DNA (deoxyribonucleic acid) of living cells.

Deoxyribonucleic acid contains the genetic material which determines what each person will be. Any interference with this material will produce mutagenic effects ranging from imperceptible defects to gross abnormalities. According to work done by Dr. Herbert S. Rozenkrantz, College of Physicians and Surgeons, Columbia University, ethylene oxide is a mutagen and possibly a carcinogen. Authorized for use in Canada and the United States.

Ethyl maltol – *See* Maltol.

Ethyl vanillin – *See* Vanillin.

Fast green (Green #3) – *See* Artificial colour.

Fat free (Low fat, Fat reduced) – The recent trend toward eating less unhealthy fat is certainly a step in the right direction to improve anybody's diet. It is also ridiculously easy to do. Remember, the idea should be to eat good, healthy fats, not to eliminate it completely from the diet, nor to replace it. By obtaining the bulk of your diet from fresh fruits, vegetables, nuts, seeds and grains and avoiding hydrogenated fats and saturated fats, we can drastically reduce the bad fat content in our diets. The unfortunate downside of the low-fat trend has been to produce yet another market for manufacturers, though not a particularly good one for consumers. With the addition of more of the existing additives and the invention of new ones such as olestra, manufacturers are besieging us with more and more unhealthy low-fat foods and snacks. The label may imply that the food is "good for you" but there is a good chance it will be detrimental to your health. Remember, if a product has a creamy texture and all the other qualities of fat yet is "fat-reduced," there has to have been something else added. And that is more additives! Leave these items on the shelf and re-learn the joys of eating simple, basic foods as Mother Nature intended.

Ferrous gluconate – may be used in ripe olives for colour retention. It is an iron salt of gluconic acid which is found naturally in the body and is sometimes used in the treatment of iron deficiency anemia. It has been known to cause gastrointestinal upsets. Authorized for use in Canada and the United States.

Ficin – may be used in ale, beer, light beer, malt liquor, porter, stout, sausage casings, meat cuts, meat-tenderizing preparations and hydrolyzed animal, milk and vegetable protein as a food enzyme. It is produced commercially from the latex of the fig tree. Ficin is also used in its concentrated state as a meat tenderizer and is ten to twenty times more potent than the tenderizer papain. Medically, it can be used in the manufacture of wound-debriding agents (agents that dissolve dead tissue to prepare the wound for healing). Ficin is capable of causing irritation to the skin, eyes and mucous membranes. Since many enzymes are now being taken from genetically altered sources, there is no guarantee that ficin will not be genetically engineered in the future also. *See* Genetically Engineered Foods, page 109. Authorized for use in Canada and the United States.

Flavour – *See* Artificial flavour.

Fructose – is commercially prepared in numerous ways and is the sweetest of the sugars. It occurs naturally in fruit and honey. Fructose is absorbed by the human digestive system more slowly than sugars such as sucrose and it causes the blood glucose level to increase only slightly after consumption. Fructose is sometimes used to improve the texture of ice cream; medically it is used as an intravenous nutrient. *See also* Sugar.

Fumaric acid – may be used in gelatin and unstandardized foods as a pH adjusting agent, acid-reacting material and/or water-correcting agent. It is important for cell metabolism in every tissue of the body. Fumaric acid is derived from numerous plants and can also be produced synthetically. Tests on humans and animals suggests that it is safe. Authorized for use in Canada and GRAS in the United States.

Furcelleran (Ammonium furcelleran, Calcium furcelleran, Potassium furcelleran, Sodium furcelleran) – may be used in ale, beer, light beer, malt liquor, porter, stout, calorie-reduced margarine, canned asparagus, green beans, waxed beans, peas and unstandardized foods as an emulsifying, gelling, stabilizing and/or thickening agent. Furcelleran is on the FDA's list of additives to be studied for mutagenic, teratogenic, subacute and reproductive effects. Since furcelleran is similar in chemical structure to carrageenan, some researchers worry that the same concerns may apply to this additive. Authorized for use in Canada and in the United States. *See also* Carrageenan.

Gelatin – may be used in cream, head cheese, jellies, meat loaf, milk (naming the flavour), mustard pickles, prepared hams, processed cheese, relishes, ice cream, sour cream, sherbet and unstandardized foods as a stabilizer and/or thickening agent. Contrary to the popular opinion that gelatin is pure protein and has no known toxicity, gelatin is loaded with chemicals. If you would like to know how Health Canada legislates the addition of chemicals in gelatin, please see Section B.14.062 of the *Food and Drugs Act*, page 251.

Gluconic acid – may be used in unstandardized foods as a pH adjusting agent, acid-reacting material and/or water-correcting agent. This is one of many organic acids which is commercially produced by a process which induces various moulds to ferment certain sugars (such as glucose). Authorized for use in Canada and GRAS in the United States.

Glucono-delta-lactone – is an acid used in cooked sausage and meat loaf to accelerate colour fixing; in dry sausage to assist in curing and in unstandardized foods as a pH adjusting agent, acid-reacting material and/or water-correcting agent. Authorized for use in Canada and the United States.

Glucose – *See* Sugar.

Glyceride (Monoglycerides, Mono- and Diglycerides) – may be used in infant formula, bread, cream, processed cheese, cocoa, milk chocolate, sweet chocolate, cottage cheese, ice cream mix, ice milk mix, non-edible sausage casings, margarine, sherbet, shortening, sour cream and unstandardized foods as an emulsifying, gelling, stabilizing and/or thickening agent; in jams and jellies, marmalade and unstandardized foods as an anti-foaming agent or release agent; in flavours, oil-soluble annatto, annatto butter colour, annatto margarine colour, unstandardized flavouring preparations and food additive preparations as a carrier or extraction solvent. According to the *Food and Drugs Act*, Division 9, Section B.09.012, glyceride may contain phosphoric acid, an anti-foaming agent and Class IV preservatives.

Mono- and diglycerides belong to a family of chemicals known as glycerides which occur naturally. However, when used as food additives, they are usually synthetically prepared with the use of many different compounds. Studies on animals have shown that different members of the glyceride family cause poor growth, high mortality, decreased ability to absorb essential fatty acids, enlarged kidneys and livers, significantly smaller testes and discolouration of the uterus. The FDA has diglyceride on its list of additives to be studied for mutagenic, teratogenic, subacute and reproductive effects. Authorized for use in Canada and GRAS in the United States.

Glycerine (Glycerol) – may be used in meat-curing compounds and sausage casings as a humectant; in preserved meats as a glaze and in unstandardized foods as a humectant and plasticizer. It is also permitted in flavour extract, flavour essence, flavour, unstandardized flavouring preparations, colouring mixtures/preparations and in food additive preparations as a carrier or extraction solvent. Glycerine can also be used in the manufacture of cough drops, face cream and suppositories. Authorized for use in Canada and the United States.

Glycerol – *See* Glycerine.

Glyceryl triacetate – *See* Triacetin.

Glyceryl tributyrate – *See* Tributyrin.

Glycine (Aminoacetic acid) – may be used in mono- and diglycerides as a sequestering agent. It is an amino acid present in all proteins. Glycine is most commonly found in gelatin. It is used medically as a gastric antacid. It was removed from the GRAS list as large quantities were proven to retard growth and increase the death rate in animals. A complete ban of glycine in special dietary sweeteners was proposed by the Food and Drug Administration. Authorized for use in Canada.

Gold – *See* Artificial colour.

Green #1 (Guinea green B) – *See* Artificial colour.

Green #2 (Light green SF yellow) – *See* Artificial colour.

Green #3 (Fast green) – *see* Artificial colour.

Guaic gum – *See* Gum guaiac.

Guar gum – may be used in infant formula, cream, salad dressings, French dressing, milk, processed and cream cheese, relishes, cottage cheese, ice cream, mustard pickles, sherbet, unstandardized foods, calorie-reduced margarine and sour cream as an emulsifying, gelling, stabilizing and/or thickening agent. It is obtained from the guar plant grown in India. After a long-term animal feeding study carried out in the US, biochemists consider it GRAS when used in amounts determined by the Food and Drug Administration. Authorized for use in Canada.

Gum Arabic (Acacia gum, Gum Senegal) – is permitted in cream, French dressing, ice cream, ale, beer, stout, porter, milk, ice milk, sherbet, calorie-reduced margarine, canned asparagus, green beans, waxed beans, peas and unstandardized foods as an emulsifying, gelling, stabilizing and/or thickening agent and in confectionery as a glazing and polishing agent. It is obtained by manufacturers from Sudanese acacia trees. Medically, gum arabic is used as an emollient to soothe irritated mucous membranes. Although its oral toxicity is considered low, some people may develop allergic reactions such as asthmatic attacks or skin rashes from gum arabic. The FDA considers this additive to have been inadequately studied and has given gum arabic priority for testing of its mutagenic, teratogenic, subacute and reproductive effects. Authorized for use in Canada and GRAS in the United States.

Gum benzoin – may be used in confectionery as a glazing and polishing agent. It is one of several resins which contain benzoic acid and is obtained from a variety of trees found in China, Thailand and Cambodia. Gum benzoin is also employed as tincture of benzoin, as a skin protector, a respiratory inhalant, an ingredient for perfumes and cosmetics. It is considered non-toxic. Authorized for use in Canada and the United States.

Gum guaiac (Guaiac gum) – may be used in fats, oils, lard, shortening and some unstandardized foods as a preservative. It is a resin obtained from the tropical guaiacum tree. Authorized for use in Canada and GRAS in the United States.

Gum Senegal – *See* Gum Arabic.

Hexane – may be used in spice extracts, natural extractives and vegetable fats and oils as a carrier or extraction solvent. This volatile liquid is obtained through the distilling of petroleum. It has been known to cause irritation to the respiratory tract and may also be a mild depressant to the central nervous system. Authorized for use in Canada and the United States. *See also* Petrolatum.

HVP – *See* Hydrolyzed vegetable protein.

Hydrochloric acid – is permitted in infant formula, ale, beer, light beer, malt liquor, porter, stout and gelatin as a pH adjusting agent, acid-reacting material and/or water-correcting agent. It is also used in starch as a modifying agent. According to Division 16, Table X of the *Food and Drugs Act*, the maximum level of use is governed by Good Manufacturing Practice. This allows manufacturers to use any amount they feel is necessary to achieve the desired effect.

Hydrochloric acid in concentrated form has caused permanent eye damage, severe burns and skin rashes. Ingestion of this additive has been known to erode the mucous membranes, esophagus and stomach. Inhaling the fumes leads to choking and inflammation of the respiratory tract. The results could be circulatory collapse and death. Authorized for use in Canada and GRAS in the United States. *See also* Starch.

Hydrogenated fat (Partially hydrogenated fat) – is the result of a chemical process in which edible oil is exposed to a high temperature and placed under pressure. Then, in the presence of nickel or platinum, hydrogen gas is bubbled through the oil, which turns it into a solid. It is also subjected to bleaching, filtering and deodorizing in order to obtain shortening, margarine and lard. The new product contains unnatural fatty acids which the body has great difficulty in handling. The most serious of these is the trans-fatty acids. Trans-fatty acids were implicated in the increase of heart disease in a study done by Dr. Walter Willett of the Harvard School of Public Health. Dr. Willett studied 60,000 nurses over a period of years and found that those nurses who consumed margarine (loaded with trans-fatty acids) had significantly more heart disease than nurses who didn't eat margarine.

During the heating process, all vitamins, minerals and essential fatty acids are destroyed. This procedure also produces a change in the oil's proteins. The residues of metals such as nickel in hydrogenated fats can pose a hazard due to their cumulative effect which could lead to metal poisoning. Recurring respiratory infections can be caused by excesses of nickel. Authorized for use in Canada and the United States.

Hydrogen peroxide – is used in starch as a modifying agent; in brewers' mash as a clarification aid and in liquid whey to decolourize and maintain pH. This chemical is a strong oxidizer which can, in concentrated solution, cause burns to the skin and mucous membranes. The FDA has it on its list to be studied for short-term mutagenic, teratogenic, subacute and reproductive effects. Authorized for use in Canada and GRAS in the United States. *See also* Starch.

Hydrolyzed vegetable protein (HVP, Protein hydrolysate) – is not considered an additive according to Division 1 of the *Food and Drugs Act*. It is used to enhance the natural flavour of certain foods. Hydrolyzed vegetable protein is made up of vegetable protein (generally soyabean) that has been broken down by a chemical process into amino acids. One of these amino acids is glutamic acid which is the main component of monosodium glutamate (MSG). Hydrolized vegetable protein may contain up to forty per cent monosodium glutamate. As MSG and HVP are flavour enhancers, they allow manufacturers to use fewer nutritious ingredients in foods.

Research on MSG has shown that every species of animal tested developed brain lesions. Infant animals were especially susceptible as they had not yet developed an effective blood-brain barrier. North American consumers learned of these dangers and refused to purchase baby foods containing monosodium glutamate. In 1969, manufacturers responded by removing it from their products. Inconceivably, they began using MSG again under the harmless sounding name hydrolyzed vegetable protein. Consumers were outraged when they learned of this deception. For the second time, manufacturers were forced to improve the quality of baby food.

It should be pointed out that the Canadian government imposed no restrictions on the use of MSG or HVP in baby foods and yet, baby food manufacturers removed both of these dangerous additives. This underlines the fact that consumer demand and decreased sales force manufacturers to make positive changes. Industry policy is to make money. It is up to the consumer, however, to remain vigilant and make industry accountable. As shown in the case of MSG and HVP, consumer outrage was effective and industry reacted to the public's refusal to buy their products.

Hydrolized vegetable protein, like MSG, has caused brain damage in infant mice. Up until 1978, HVP comprised up to two per cent of the weight of some baby foods. Due to the dangers of these additives, consumers would be well advised to avoid products containing HVP or monosodium glutamate. Authorized for use in Canada and the United States. *See also* Monosodium glutamate.

Hydroxylated lecithin – may be used in cocoa, milk chocolate, sweet chocolate and unstandardized foods as an emulsifying, gelling, stabilizing and/or thickening agent. It is produced by a process in which a compound is made more soluble by the introduction of an atom of hydrogen and an atom of oxygen.

Although the FDA has cleared it for use as a food emulsifier, some scientists suggest that the public should avoid the use of this chemical. The FAO/WHO asserts the safety of hydroxylated lecithin has not been adequately established. Authorized for use in Canada and the United States.

Hydroxypropyl cellulose – may be used in unstandardized foods as an emulsifying, gelling, stabilizing and/or thickening agent, and is governed by the Good Manufacturing Practice regulation. Authorized for use in Canada. *See also* Sodium carboxymethylcellulose.

Hydroxypropyl methyl cellulose – may be used in salad dressings, milk, mustard pickles, relish, skim milk, partly skim milk and unstandardized foods as an emulsifying, gelling, stabilizing and/or thickening agent. It is governed by the regulation Good Manufacturing Practice. Authorized for use in Canada. *See also* Sodium carboxymethylcellulose.

Indigotine (Blue #2) – *See* Artificial colour.

Invert sugar – is considered an ingredient, rather than an additive, according to Division 1 of the *Food and Drugs Act*. It is produced by combining equal amounts of glucose and fructose. (*See also* Sugar.) In the US, it is used in confectionery and in brewing to retain moisture. Medically it is also used in intravenous solutions. In 1980, the final report to the FDA of the Select Committee on GRAS Substances stated that there is no evidence that invert sugar is a hazard. Authorized for use in Canada and GRAS in the United States.

Iron oxide (Rust) – *See* Artificial colour.

Isoascorbic acid – *See* Erythorbic acid.

Isolated soy protein – *See* Textured vegetable protein.

Isopropyl citrate – *See* Stearyl citrate.

Kadaya – *See* Karaya gum.

Karaya gum (Kadaya) – may be used in French dressing, salad dressing, milk, cream cheese, cottage cheese, ice cream, sherbet, relishes, mustard pickles, unstandardized foods and calorie-reduced margarine as an emulsifying, gelling, stabilizing and/or thickening agent. It is a carbohydrate which is obtained from a tree in India. It has been linked to allergic responses and should be tested further for other possible adverse effects. Authorized for use in Canada and GRAS in the United States.

Lactase – can be used in milk destined for use in ice cream mixes, flour, bread, yogurt and whey as a food enzyme. It is obtained commercially from various yeasts. It has no known toxicity. Since many enzymes are now being taken from genetically altered sources, there is no guarantee that lactase will not be genetically engineered also. For more detailed information on genetically altered foods and additives, please see Genetically Engineered Foods, page 109. Authorized for use in Canada and GRAS in the United States.

Lactic acid – may be used in bread, cider, cottage cheese, egg whites and yolks, salad dressing, ice cream mix, mayonnaise, olives, pickles, relishes, processed cheese, sherbet, margarine, beer, light beer, stout, porter, unstandardized foods, canned pears and canned strawberries as a pH adjusting agent, acid-reacting material and/or water-correcting agent. Lactic acid occurs naturally in sour milk, molasses and certain plants and is manufactured commercially by the fermentation of some of these products. Authorized for use in Canada and GRAS in the United States.

Lactose – is not considered an additive according to Division 1 of the *Food and Drugs Act*. It is present only in the milk of mammals and is said to be virtually safe, except for those suffering from a genetic disease called galactosemia. It is also produced synthetically. Authorized for use in Canada and the United States.

Lanolin – may be used in chewing gum as a plasticizing agent. It is also employed as a base in some cosmetics. Lanolin is a secretion from the sebaceous glands of sheep. This additive has been known to cause allergic reactions in some people. Authorized for use in Canada and the United States.

Larch gum – *See* Arabinogalactan.

Lead – Although not a food additive as such, residual limits are permitted at certain levels in infant formula (those that are ready-to-serve), evaporated milk and condensed milk.

In their forty-first report, the FAO/WHO Expert Committee on Food Additives, 1993, assessed the risks posed to infants and children who consumed lead. The most significant effect was the reduction of cognitive development and intellectual performance. Children who consumed small amounts of lead showed a decrease of one to three IQ points for each 10 microgram per decilitre increment of lead in the blood. It is known that lead crosses from the mother to the fetus via the placenta. Existing studies suggest there is no evidence of a threshold for lead. *See* B.15.001, Division 15, page 253 to see how the government legislates acceptable levels of lead in foods such as infant formula, evaporated milk and/or condensed milk. At their forty-fourth meeting, the FAO/WHO Expert Committee on Food Additives, 1995, recommended that in future, submissions of substances under review should indicate actual concentrations of lead.

Lecithin – may be used in infant formula, bread, cream, milk, relishes, processed and cream cheeses, cocoa, ice cream, milk chocolate, mustard pickles, sweet chocolate, sherbet, margarine and unstandardized foods as an emulsifying, gelling, stabilizing and/or thickening agent and in fats, oils, lard, mono- and diglycerides, shortenings and some unstandardized foods as a preservative. It occurs naturally in almost all plant and animal tissues and is generally obtained commercially from soyabeans, corn and eggs. Lecithin is considered to be non-toxic. Authorized for use in Canada and GRAS in the United States.

Licorice (Licorice extract, Licorice extract powder, Licorice root) – is not considered an additive according to Division 1 of the *Food and Drugs Act*. It is obtained from a Mediterranean plant and is used as a flavouring in numerous products such as candy, licorice and certain beverages. Numerous cases have been reported where individuals who ate as little as 228 grams (eight ounces) of licorice daily, for prolonged periods, developed elevated blood pressure, headaches, shortness of breath, swelling and paralysis of the extremities and heart palpitations. It is presently on the FDA's list for further testing. Authorized for use in Canada and GRAS in the United States. *See also* Ammoniated glycyrrhizin.

Locust bean gum – *See* Carob bean gum.

Low fat – *See* Fat free.

Magnesium aluminum silicate – *See* Silicate.

Magnesium citrate – may be used in soft drinks as a pH adjusting agent, acid-reacting material and/or water-correcting agent. It can be used medically to induce vomiting or diarrhea in the treatment of certain conditions. Authorized for use in Canada.

Magnesium hydroxide – may be used in chocolate, cocoa, ice cream mix, ice milk mix, milk chocolate, sweet chocolate, canned peas and in bacterial cultures as a pH adjusting agent, acid-reacting material and/or water-correcting agent. Medically, it is used as a gastric antacid and can be used to induce vomiting or diarrhea in the treatment of certain conditions. Authorized for use in Canada and GRAS in the United States.

Magnesium oxide – may be used in certain unstandardized dry mixes as an anti-caking agent; in ice cream mix and in ice milk mix as a pH adjusting agent, acid-reacting material and/or water-correcting agent. It can be used medically as a gastric antacid. Authorized for use in Canada and GRAS in the United States.

Magnesium silicate – may be used in salt, garlic salt, onion salt, icing sugar and unstandardized dry mixes as an anti-caking agent; in confectionery as a glazing or polishing agent; in chewing gum as a dusting agent and in rice as a coating agent. Due to reports that dogs suffer kidney damage when they ingest magnesium silicate, the WHO Food Committee recommends further study of magnesium silicate. Authorized for use in Canada and GRAS in the United States. *See also* Silicate.

Magnesium stearate – may be used in salt, garlic salt, onion salt, icing sugar and unstandardized dry mixes as an anti-caking agent and in confectionery as a releasing agent. It has been used in baby dusting powders and as a tablet lubricant. Authorized for use in Canada and the United States.

Magnesium sulphate – may be used in ale, beer, light beer, malt liquor, porter, stout and bacterial cultures as a pH adjusting agent, acid-reacting material and/or water-correcting agent and in starch as a modifying agent. It has been used industrially in the manufacture of mother-of-pearl, frosted paper, fire-proofing fabrics, fertilizers, explosives and matches. It can be used medically to induce vomiting or diarrhea in the treatment of certain conditions. When injected or used by a person suffering from renal insufficiency, it may cause magnesium poisoning. Authorized for use in Canada and GRAS in the United States. *See also* Starch.

Malic acid (Apple acid) – may be used in jams, jellies, canned applesauce, canned pears, canned strawberries and unstandardized foods as a pH adjusting agent, acid-reacting material and/or water-correcting agent. This colourless crystalline compound is present in large quantities in apples and exists in all living cells. It should not be added to baby foods as it is unknown whether infants are able to metabolize it as well as adults. Authorized for use in Canada and GRAS in the United States.

Malt extract – is not considered an additive according to Division 1 of the *Food and Drugs Act*. It is obtained from germinated barley and is used as a food nutrient. It is also used extensively in the brewing industry. At present, it is not considered toxic. Authorized for use in Canada and GRAS in the United States.

Malto-dextrin – is not considered an additive according to Division 1 of the *Food and Drugs Act*. It is obtained by hydrolysing starch and combining maltol and dextrin. It is used mainly as a texturizer and flavour intensifier in candies. It has no known toxicity. Authorized for use in Canada and GRAS in the United States.

Maltol (Ethyl maltol) – is not considered an additive according to Division 1 of the *Food and Drugs Act* in spite of the fact that it is synthetically produced and is used in food to enhance flavour. Scientists have found that this chemical causes a drop in the level of hemoglobin in the blood and an increase in the level of hemosiderin (a substance produced from the breakdown of the red blood cells). The implications of this are serious enough in a healthy individual but could be disastrous for a person suffering from inherited anemia (thalassemia). It has been recommended for extensive carcinogenic and teratogenic studies. Authorized for use in Canada and the United States.

Maltose – *See* Sugar.

Manganese sulphate – may be used in ale, beer, light beer, malt liquor, porter and stout as a yeast food. This additive is governed under the Good Manufacturing Practice regulation. It is used as a commercial dye and is a component of fertilizers for vines, tobacco and feeds. Manganese occurs naturally in many minerals. It is toxic in humans when inhaled as manganese sulphate dust or fumes. The symptoms of manganese sulphate poisoning are similar to those of Parkinson's disease, e.g. lassitude, weakness, emotional disturbances, spastic gait and paralysis. Authorized for use in Canada and the United States.

Mannitol – is used in unstandardized foods as a sweetener. Although it occurs naturally in a wide variety of plants, it is synthesized from the sugars of the manna ash tree and certain seaweeds. Mannitol is used commercially in the manufacture of artificial resins and plasticizers and when combined with boric acid it is used in dry electrolytic condensers for radio applications. Medically, it is employed as a diuretic, as a diagnostic aid to test kidney function and in the preparation of antacid tablets, cough-cold tablets and children's Aspirin™. Injected intravenously, it has been implicated in a wide range of disease processes, focalizing in disturbed metabolism in humans. Other adverse effects were acidosis, dehydration, kidney stones, kidney failure, nausea, disorientation, unconsciousness and death. Authorized for use in Canada and the United States.

MC (Microcrystalline Cellulose) – *See* Methyl cellulose and Sodium carboxymethylcellulose.

Methyl cellulose (Microcrystalline methyl cellulose, MC, Microcrystalline cellulose) – may be used in salad dressings, processed cheese, ale, beer, light beer, malt liquor, porter, stout and unstandardized foods as an emulsifying, gelling, stabilizing and/or thickening agent. Its use in most of these products is governed by the Good Manufacturing Practice regulation. Methyl cellulose is a chemically modified product obtained from wood pulp. Manufacturers claim that it adds five times the amount of fiber that one hundred per cent whole wheat bread would normally contain. The related additive microcrystalline methyl cellulose is absorbed into the bloodstream and conveyed to the kidneys for excretion. Unfortunately, our kidneys may not be capable of handling an indigestible substance such as this. It has also been pointed out that excesses of the wrong type of fibre lead to the body being depleted of its nutrients. Even more serious is the possibility of portions of the large bowel becoming twisted or blocked. Authorized for use in Canada and the United States. *See also* Sodium carboxymethylcellulose.

Methyl ethyl cellulose – may be used in unstandardized foods as an aerating agent and as an emulsifying, gelling, stabilizing and/or thickening agent. Authorized for use in Canada and the United States. *See also* Sodium carboxymethylcellulose.

Methylene chloride – *See* Dichloromethane.

Methyl paraben (Methyl-p-hydroxy-benzoate) – may be used in jams, jellies, packaged fish and meat, mincemeat, pickles, relishes, tomato catsup, tomato paste, tomato purée and certain unstandardized foods as a Class II preservative. It is also used in cosmetics. This additive has been known to cause allergic skin irritations. Testing indicates that this compound suppresses the conveyance of molecules into the cells which hinders human cell growth. Further studies on cells from human livers, intestines and chick embryos showed that antimicrobial (anti-bacteria) agents such as methyl paraben were as toxic to the cells as they were to the bacteria against which they were being employed. Researchers have decided that a healthy human digestive system, under normal circumstances, may be able to detoxify these agents. Consumption of large amounts (due to widespread use of preservatives), however, may place too great a burden on the digestive system and in all likelihood, these agents would cause serious damage. The FDA has methyl paraben on its list for further study. Authorized for use in Canada and the United States.

Methyl-p-hydroxy-benzoate – *See* Methyl paraben.

Methyl polysilicone – *See* Dimethylpolysiloxane.

Methyl silicone – *See* Dimethylpolysiloxane.

Microcrystalline cellulose (MC, Cellulose microcrystalline) – may be used in ice cream mix, ice milk mix, sherbet, whipped vegetable oil topping, unstandardized frozen desserts, unstandardized sandwich spread and dips and some unstandardized foods as a filler. Authorized for use in Canada. *See also* Methyl cellulose and Sodium carboxymethylcellulose.

Mineral oil – may be used in confectionery as a glazing and polishing agent; in bakery products and seeded raisins as a releasing agent; on fresh fruit and vegetables as a coating and in edible collagen sausage casing as a lubricant. It is employed medically as a laxative, lubricant or stool softener. Mineral oil is a nonfood fat distilled from petroleum that should not be taken immediately before or after a meal as it will absorb and carry out fat-soluble vitamins such as vitamin A or beta carotene and vitamin K. Its use in some foods will obviously reduce their nutritional value, which could lead to various deficiency-related problems.

Pregnant women should take care to avoid the use of mineral oil as it could make the baby more susceptible to hemorrhagic disease. In 1991, the thirty-seventh report of the FAO/WHO Expert Committee on Food Additives evaluated studies which showed haematological changes and deposition of mineral oil in the liver, spleen and lymph nodes. The Committee decided that an adequate long-term feeding study should be performed. In 1995, at their forty-fourth meeting, the Committee reviewed studies which showed that mineral-oil-induced lesions, similar to those seen in rats, have been identified in human tissues. The studies requested in 1991 were still not available and the temporary Acceptable Daily Intake (ADI) "not specified" was extended. The Committee decided it required information about the components of mineral oil that influence its absorption and toxicology, together with all relevant background data for review in 1998. Authorized for use in Canada and the United States.

Modified starch – *See* Starch.

Monoammonium glycyrrhizin – *See* Ammoniated glycyrrhizin.

Monoammonium phosphate – *See* Ammonium phosphate, monobasic.

Mono- and di-glycerides – *See* Glyceride.

Monocalcium phosphate – *See* Calcium phosphate, monobasic.

Monoglyceride citrate – may be used in fats, oils, lard, margarine, shortening and some unstandardized foods as a preservative, and in spice extracts, natural extractives and unstandardized flavouring preparations as a carrier or extraction solvent. Authorized for use in Canada and GRAS in the United States. *See also* Calcium citrate.

Monoglyceride – *See* Glyceride.

Mono-potassium glutamate – *See* Monosodium glutamate.

Monopotassium phosphate – *See* Potassium phosphate, monobasic.

Monosodium glutamate (MSG, Mono-ammonium glutamate, Mono-potassium glutamate) – is used in thousands of processed foods in Canada and the United States. Originally, monosodium glutamate was extracted from seaweed or soyabeans. Now it is also manufactured by a chemical process from wheat or corn gluten or from sugar beet by-products.

The Canadian government does not consider MSG a food additive but labels it a "flavour enhancer," for which there are no specific regulations to govern its use (*See* B.01.009, Division 1, page 154). Consumers will find that it is nearly impossible to avoid.

Monosodium glutamate is made up of sodium and glutamic acid (an amino acid). Amino acids are the building blocks of protein and they are vital to every living organism. When glutamic acid is withdrawn from its natural food source, without its companion amino acids, it has no nutritional value whatsoever and may cause symptoms characteristic of mild poisoning because the body may attempt to handle it as it would a foreign substance. In addition, wheat, corn and sugar beet by-products, which are three of MSG's main sources, are also known allergens.

Monosodium glutamate has been known to cause what is often referred to as "Chinese Restaurant Syndrome", which is characterized by a burning sensation in the back of the neck and in the forearms, tightness of the chest and headaches. This syndrome obtained its name from the fact that Chinese restaurants are notorious for their generous use of MSG to enhance the flavour of their foods. Overuse of MSG, however, is certainly not confined to Chinese restaurants.

In laboratory experiments, every species of animal fed MSG developed brain lesions and infant animals were especially susceptible as they had not yet developed an effective blood-brain barrier. An amount of MSG comparable to that which would have been found in five 113-gram (four-ounce) jars of baby food produced nerve-cell death in seven out of nine baby animals. Adult animals treated with MSG displayed arrested skeletal development, marked obesity and sterility in the females. In other laboratory experiments, additional adverse effects were noted in the retina of the eye and in the liver. Although MSG was removed from baby food in 1969, it was reinstated under the guise of hydrolyzed vegetable protein (HVP), which contains up to forty per cent monosodium glutamate. As this information became public knowledge, the baby food industry decided to remove HVP from their products in 1978. This shows the power of consumer action!

Around 1969, scientists found that MSG seems to be a teratogen (a cause of birth defects). Scientists suggest that MSG may cause deformities at any time during the pregnancy. We encourage the public to refrain from purchasing any product that contains MSG in order to compel food manufacturers to eliminate this chemical and upgrade their products. The recent trend by many restaurants to advertise MSG-free products is cause for applause. Authorized for use in Canada and GRAS in the United States. *See also* Hydrolyzed Vegetable Protein.

Monosodium phosphate – *See* Sodium phosphate, monobasic.

MSG – *See* Monosodium glutamate.

Natural colour – *See* Artificial colour.

Natural flavour – *See* Artificial flavour.

N-butane (Butane, Methylsulphonal, Bioxiran, Dibutadiene dioxide) – may be used on edible, vegetable oil-based or lecithin-based pan coatings as a propellant. This flammable gas derived from petroleum is used as a solvent, refrigerant and food additive. The National Institute of Occupational Safety and Health has ascertained that n-butane is an animal carcinogen. Authorized for use in Canada and GRAS in the United States.

Nitrate – *See* Sodium nitrite.

Nitric acid – *See* Starch.

Nitrogen – may be used in unstandardized foods as a pressure-dispensing agent. Industrially, it is employed in the manufacture of ammonia, cyanides and explosives. Authorized for use in Canada and GRAS in the United States.

Nitrous oxide – may be used in unstandardized foods as a pressure-dispensing agent. Nitrous oxide (also known as laughing gas) has been used medically as an anesthetic. Authorized for use in Canada and GRAS in the United States.

Oat gum – may be used in unstandardized foods as an emulsifying, gelling, stabilizing and/or thickening agent. Diarrhea, intestinal gas and allergic reactions have been known to occur occasionally as a result of eating foods containing oat gum. It has no known toxic effects. Authorized for use in Canada and the United States.

Octenyl succinic anhydride – *See* Starch.

Olean™ – *See* Olestra.

Olestra (Olean™) – is a "fake" fat recently approved by the FDA for use in low-fat foods. It is produced by combining a sucrose (sugar) molecule with up to eight fatty acids taken from a variety of vegetable oils such as soyabean, corn, palm, coconut and cottonseed. The resulting olestra molecules are so large and fatty that the naturally occurring enzymes and bacteria in the intestines are unable to metabolize them. While molecules that pass through the body without being absorbed should in theory, be safe, the olestra molecules raise the following concerns:

- (Since the gastrointestinal system does not absorb olestra molecules,) the possible adverse effects on the gastrointestinal system as olestra passes through it are impossible to know without extensive testing.
- The olestra molecules could interfere with the absorption of fat-soluble vitamins, other nutrients and drugs by carrying them out of the body before they can be metabolized.

Olestra was created in 1968 by researchers at Procter and Gamble. They were looking for a product that could be used to increase fat intake in premature babies. Due to olestra's indigestibility, this use was not feasible. Then, in 1975, Procter and Gamble sought approval for its use as a drug as they believed it would be effective in lowering blood cholesterol levels. Studies showed that olestra did not perform well enough in this regard to warrant its use as a drug.

In 1987, Procter and Gamble petitioned the FDA once again for permission to use olestra, this time seeking to replace some or all of the fat in shortenings and cooking oils, fast foods, chips and other products. In 1990, after the approval was held up due to the results of testing and pressure from concerned groups such as the Center for Science in the Public Interest (CSPI), Procter and Gamble petitioned the FDA once more. This time the company was seeking approval to use olestra in the more limited capacity of replacing fat in "savory snacks" such as potato chips, cheese puffs and crackers. Finally in January 1996, the FDA granted approval for olestra to be used in savory snacks (such as chips, crackers and tortilla chips). Although it was approved for use, the FDA required products to carry the following declaration on the label:

> This product contains olestra. Olestra may cause abdominal cramping and loose stools. Olestra inhibits the absorption of some vitamins and other nutrients. Vitamins A, D, E and K have been added.

In 1996 and 1997, Frito-Lay and Procter and Gamble began test marketing some olestra snacks in certain US states. By mid 1997, CSPI presented the FDA with over one thousand reports from consumers who had suffered ill effects from these products. The effects reported were the same as those experienced by laboratory animals and included severe abdominal cramping and uncontrollably loose stools. In the case of children, the elderly or chronically ill, these symptoms could prove dangerous.

As a note of interest, Proctor and Gamble insisted that there was no proof that olestra had caused these problems and that these cases were just coincidences. It is hard to believe that it was a coincidence that so many people suffered symptoms identical to the company's own test animals!

Despite these problems, Frito-Lay and Procter and Gamble plan to go ahead and market olestra snacks in the United States. The FDA intends to re-evaluate olestra's safety after thirty months.

The commercials that we have seen for olestra (Olean™) depict it as a modern, nutritious product that will help us to "get the fat out of our diet." The advertisements depict earthy farmers and their wives tending beautiful fields of soyabeans and expressing their pleasure at being involved in the production of this new product.

In the meantime, Procter and Gamble will continue to seek permission to use olestra in products in Canada and the United Kingdom.

Orange B – *See* Artificial colour.

Orange #1 – *See* Artificial colour.

Orange #2 – *See* Artificial colour.

Orange oil – is not considered an additive according to Division 1 of the *Food and Drugs Act*. It is an extremely yellow or orange liquid used as a flavouring. It is extracted from the peel of certain sweet, ripened fruit, such as oranges. Although orange oil is not considered to be toxic, the possibility of the fruit's peel being contaminated by the spraying of insecticides, herbicides and fungicides hasn't even been considered. Most oranges have been dyed with citrus red #2 to maintain a uniform colour. This dye has caused cancer in test animals. The WHO has recommended that red #2 not be used on food. Orange oil is GRAS in the United States. Authorized for use in Canada. *See also* Artificial colour.

Oxystearin – may be used in cottonseed, peanut and soyabean oils to inhibit crystal formation. It is also used in the manufacture of cosmetics, soaps, candles, pill coatings and suppositories. Oxystearin is a modified fatty acid that occurs in animal fat. It has been suggested by the Select Committee of the Federation of American Societies for Experimental Biology that further study is necessary on this additive. Authorized for use in Canada and the United States.

Papain – may be used in beer, beef before slaughter, edible collagen sausage casings, hydrolyzed animal, milk and vegetable protein, meat cuts, meat-tenderizing preparations and precooked or instant cereals as a food enzyme. According to the *Food and Drugs Act*, the only commercially permitted source for papain is the fruit of the papaya plant. It is added to enriched farina which reduces its cooking time. The papain used in food can digest approximately thirty-five times its own weight of lean meat; if it can do this to meat just think of what it can do to your digestive tract! It has been known to cause allergic reactions, in varying degrees, in susceptible individuals. Since many enzymes are now being taken from genetically altered sources, there is no guarantee that papain will not be genetically engineered also.

Please see Genetically Engineered Foods, page 109. Authorized for use in Canada and the United States.

Paprika – may be used in bread, butter, concentrated fruit juice (unfrozen varieties), ice cream mix, icing sugar, jam with pectin, liqueurs, milk, pickles, relishes, sherbet, smoked fish, lobster paste, caviar, tomato catsup, liquid, dried or frozen whole egg or egg yolk, vegetable fats and oils, margarine, a variety of cheeses and unstandardized foods as a colouring agent. It is obtained from the ground pods of the sweet pepper. Authorized for use in Canada and the United States.

Paraffin – *See* Petrolatum.

Paraffin wax – may be used on fresh fruit, vegetables and cheese as a coating. It is derived from petroleum. The intestines do not metabolize or absorb this additive. Industrially, it is employed in wax paper, candles, ointments and many other products. It is used medically in wound dressings and cosmetic surgery.

According to the thirty-ninth report of the FAO/WHO Expert Committee on Food Additives, 1992, it was decided that, for newer formulations of paraffin wax, new specifications were required and that adequate long-term studies on its mutagenic, reproductive and teratogenic properties should be completed. Authorized for use in Canada and the United States.

Partially hydrogenated fat – *see* Hydrogenated fat.

Pectin – may be used in ice cream, sherbet, jams, jellies, cream, salad dressings, milk, relishes, sour cream and unstandardized foods as an emulsifying, gelling, stabilizing and/or thickening agent. These qualities make it useful in cosmetics and drugs. This additive occurs naturally in ripe fruit and is not considered toxic. Authorized for use in Canada and the United States.

Pepsin – may be used in cheese, cottage cheese, cream cheese, precooked cereals, defatted soya flour, ale, beer, porter, stout, and malt liquor as a food enzyme. Manufacturers extract it from the glandular layer of porcine stomach (stomach of pigs). It is an enzyme secreted in the human stomach to aid in the digestion of food, particularly animal protein. It is not considered to be toxic. However, due to the similarity between meat and the linings of the stomach and intestines, high concentrations of pepsin in the digestive tract tend to cause breakdown of the linings and formation of ulcers. It has been observed that the levels of pepsin are significantly lower in a vegetarian's diet. Since many enzymes are now being taken from genetically altered sources, there is no guarantee that pepsin will not be genetically altered also. For information on genetically engineered enzymes, see Table V, page 287. Authorized for use in Canada and the United States.

Peracetic acid (Peroxyacetic acid) – may be used in starch as a modifying agent. It is obtained from acetaldehyde and is comprised of forty per cent acetic acid, which is highly corrosive. When heated to 110°F (43°C), it explodes violently. It is permitted for use in starch in the US and is governed by the Good Manufacturing Practice regulation in Canada. *See also* Starch.

Peroxyacetic acid – *See* Peracetic acid.

Petrolatum (Petroleum, Paraffin jelly) – may be used in or upon confectionery as a glazing and polishing agent; on fresh fruit and vegetables as a coating. Medically, it is used as a protective dressing and is known as Vaseline™. It has a mild laxative effect and although it is not absorbed by the body, it may interfere with digestion and the absorption of essential nutrients. It has been suggested that products of petroleum "cracking" may be cancer-inciters (substances that on their own may not cause cancer but may contribute to the development of cancer). Cracking is the process by which crude oil is broken down into different products such as petroleum jelly and paraffin wax. Authorized for use in Canada and the United States.

Petroleum – *See* Petrolatum.

Phenylalanine – *See* Aspartame.

Phosphoric acid – may be used in ale, beer, light beer, stout, porter, malt liquor, chocolate, cocoa, cottage cheese, gelatin, milk chocolate, sweet chocolate, mono- and diglycerides, as a pH adjusting agent, acid-reacting material and/or water-correcting agent. It is obtained commercially from phosphate rock and is employed in the manufacture of fertilizers, detergents and in the purification of hydrogen peroxide. It is also used in dental cements and in the rust-proofing of metals. Phosphoric acid, in concentrated solutions, is irritating to the skin and mucous membranes.

Although the body needs phosphoric acid, too much is not a good thing. It can disturb the absorption of minerals such as calcium from the gastrointestinal tract. Phosphate is a component of a great many food additives and so the body can be easily overloaded with phosphates.

Experiments performed in Germany in 1957 by Dr. G. J. von Esch and his colleagues revealed that rats that were fed a five per cent phosphate diet exhibited arrested development, diminished fertility and decreased lifespan. Authorized for use in Canada and GRAS in the United States.

Phosphorous oxychloride – may be used in starch as a modifying agent. Industrially, it is employed as a chlorinating agent and as a solvent in cryoscopy (study of fluids and their freezing points). It is a strong irritant to the skin, eyes and mucous membranes. Inhalation of phosphorus oxychloride fumes may cause pulmonary edema. Researchers have suggested that the safety of this chemical is questionable. Authorized for use in Canada and GRAS in the United States. *See also* Starch.

Polyglycerol ester (of Fatty acids) – may be used in vegetable oils, calorie-reduced margarine and unstandardized foods as an emulsifying, gelling, stabilizing and/or thickening agent. It is manufactured from edible fats, oils and fatty acids of corn, cottonseed, peanut, sesame, soyabean oils as well as lard and tallow. It is considered non-toxic. Authorized for use in Canada and the United States.

Polyoxyethylene (20) sorbitan monooleate – *See* Polysorbate 80.

Polyoxyethylene (20) sorbitan monostearate – *See* Polysorbate 60.

Polyoxyethylene (20) sorbitan tristearate – *See* Polysorbate 65.

Polyoxyethylene (8) stearate – may be used in unstandardized bakery foods as an emulsifying, gelling, stabilizing and/or thickening agent. It is a mixture of stearate and ethylene oxide. It is a waxy solid that is added to bakery products to make them feel fresh. Experiments have shown that when rats were fed a twenty-five per cent diet of polyoxyethylene (8) stearate, they developed bladder stones and numerous tumours. It was banned in the US in 1952. Authorized for use in Canada.

Polysorbate 60 (Polyoxyethylene (20) sorbitan monostearate) – may be used in cake icing, cake icing mix, puddings, pie fillings, beverage base or mix, sour cream substitute, dry soup base or mix, dry batter coating mixes, prepared alcoholic cocktails and unstandardized spreads and dips as an emulsifying, gelling, stabilizing and/or thickening agent. There are only slight technical differences between the polysorbates.

There has been a great deal of controversy among scientists regarding the polysorbates since the 1950s. Some scientists feel the polysorbates are safe even though experiments on a variety of test animals have indicated changes in tissues, sex organs, liver and kidneys, as well as tumours. In addition, the poly derivatives have caused excessive amounts of iron to be absorbed from common foods in the intestines of laboratory animals. This iron is then stored in organs such as the liver where it causes cirrhosis. In 1956, it was also suggested by the FAO/WHO Expert Committee and the International Union on Additives that some of the polysorbates were "cancer-inciters". The FDA requested further studies to be undertaken on two out of three polysorbates (polysorbate 60 and polysorbate 80). Authorized for use in Canada and the United States.

Polysorbate 65 (Polyoxyethylene (20) sorbitan tristearate) – may be used in milk, ice cream, ice milk, sherbet, cakes, unstandardized confectionery coatings, beverage base or mix, breath freshener products in candy, tablet or gum form and unstandardized frozen desserts as an emulsifying, gelling, stabilizing and/or thickening agent. Authorized for use in Canada and the United States. *See also* Polysorbate 60.

Polysorbate 80 (Polyoxyethylene (20) sorbitan monooleate) – may be used in ice cream, ice cream mix, ice milk, sherbet, pickles, relishes, beverage base or mix, imitation dry cream mix, cake icing, salt, whipped cream, breath freshener products, creamed cottage cheese and unstandardized frozen desserts as an emulsifying, gelling, stabilizing and/or thickening agent. Authorized for use in Canada and the United States. *See also* Polysorbate 60.

Polyvinylpyrrolidone – may be used in beer, cider, malt liquor, porter, stout, wine as a clarifying agent; in table-top sweetener tablets containing aspartame as a tablet binder; in colour lake dispersions for use in confectionery in tablet form as a viscosity-reduction agent (an agent that reduces the flow of a liquid) and stabilizer. When used medically as a plasma expander, it has been known to cause liver and kidney damage. The body may take up to a year to excrete it, which is a cause for concern considering it has caused cancer in test animals. It has also induced spontaneous miscarriage in test animals. Authorized for use in Canada and the United States.

Ponceau SX (Red #4) – *See* Artificial colour.

Potassium acid tartrate – may be used in baking powder, honey wine and unstandardized foods as a pH adjusting agent, acid-reacting material and/or water-correcting agent. It is a salt of tartaric acid and is considered to be non-toxic. Authorized for use in Canada and the United States.

Potassium Alginate – *See* Alginate.

Potassium aluminum sulphate (Alum, Aluminum potassium sulphate) – may be used in pickles, relishes and unstandardized foods as a firming agent; in flour and whole wheat flour as a carrier of benzoyl peroxide; in ale, beer, light beer, porter, stout, malt liquor, baking powder, oil-soluble annatto and unstandardized foods as a pH adjusting agent, acid-reacting material and/or water-correcting agent. It is employed industrially in sizing paper and waterproofing fabrics. Medically, it can be used as an astringent. In experiments on animals, it has caused gum tissues to break down, kidney damage and bleeding in the intestines. Consuming thirty grams (one ounce) can cause death in adult humans. Authorized for use in Canada and the United States.

Potassium bisulphite – may be used in ale, beer, light beer, porter, stout, malt liquor, honey wine, wine, cider, jams, jellies, pickles, relishes, refiners' molasses, table molasses, tomato catsup, tomato paste, tomato purée, beverages, dried fruit, dried vegetables, frozen mushrooms, glucose, glucose syrup, dextrose and some unstandardized foods as a Class II preservative. It is a salt of tartaric acid. Authorized for use in Canada and the United States. *See also* Sulphite.

Potassium bromate – may be used in flour, whole wheat flour, bread and unstandardized bakery foods as a bleaching, maturing and/or dough-conditioning agent.

In a review of potassium bromate at the forty-fourth meeting of the FAO/WHO Expert Committee on Food Additives, delegates endorsed the recommendation that bromate should not be present in food. This would also apply to other uses of potassium bromate in food processing such as in treating barley and brewing beer. Testing had shown it produced cancer in the kidneys, stomach and thyroid of rats. Given that potassium bromate is genotoxic and carcinogenic and residues have been detected in bread, the committee concluded that the use of bromate in flour was not appropriate. In 1958, the use of chemical oxidizing agents such as potassium bromate, was banned in Germany. Authorized for use in Canada and the United States.

Potassium carbonate – may be used in chocolate, cocoa, milk chocolate, sweet chocolate, margarine, cream cheese spreads, processed cheeses and unstandardized foods as a pH adjusting agent, acid-reacting material and/or water-correcting agent. This chemical is odourless, has a strong alkaline taste and is an irritant with caustic action. Authorized for use in Canada and the United States.

Potassium carrageenan – *See* Carrageenan.

Potassium chloride – may be used in ale, beer, light beer, malt liquor, porter, stout and unstandardized foods as a pH adjusting agent, acid-reacting material and/or water-correcting agent. When taken orally it may irritate the gastrointestinal tract (vomiting, diarrhea and ulcer formation) and cause weakness and shock. In low-sodium diet foods, it is sometimes used as a substitute for sodium chloride. Authorized for use in Canada and GRAS in the United States.

Potassium citrate – may be used in ale, beer, light beer, malt liquor, porter, stout and unstandardized foods as a pH adjusting agent, acid-reacting material and/or water-correcting agent. Medically, it is used as a gastric antacid and urinary alkalizer. It is considered non-toxic. Authorized for use in Canada and GRAS in the United States.

Potassium cyclamate – *See* Cyclamate.

Potassium furcelleran – *See* Furcelleran.

Potassium hydroxide – may be used in infant formula, chocolate, cocoa, milk chocolate, sweet chocolate, ice cream mix, ice milk mix, grape juice and unstandardized foods as a pH adjusting agent, acid-reacting material and/or water-correcting agent. It is an extremely corrosive alkali. When taken internally it has been known to produce severe pain in the throat, hemorrhaging and collapse, possibly leading to stricture of the esophagus. It is interesting to note that it has also been used to inhibit the growth of horns in calves.

The FDA has restricted the amount of potassium hydroxide to less than ten per cent in household products while at the same time deeming it GRAS for use in food. In spite of the dangers, the Canadian government has given the food industry complete discretion over the levels of potassium hydroxide it puts in food as most of the above-mentioned products are governed by the Good Manufacturing Practice regulation, most notably infant formula. This is irresponsible and unconscionable – particularly where infant formula is concerned. Authorized for use in Canada and GRAS in the United States.

Potassium iodate – may be used in bread and unstandardized bakery foods as a bleaching, maturing and/or dough-conditioning agent.

Experiments on dogs showed that the ingestion of this chemical caused irritation to the gastrointestinal tract and anemia. In test animals, death was preceded by a loss of appetite and severe weakness. In 1958, the use of chemical-oxidizing agents, such as potassium iodate, was banned in Germany. It is still authorized for use in Canada and GRAS in the United States.

Potassium metabisulphite – may be used in ale, beer, light beer, porter, stout, malt liquor, honey wine, wine, cider, jams, jellies, pickles, relishes, refiners' molasses, table molasses, tomato catsup, tomato paste, tomato purée, beverages, dried fruit, dried vegetables, frozen mushrooms, glucose, glucose syrup, dextrose and some unstandardized foods as a Class II preservative. Authorized for use in Canada and the United States. *see also* Sulphite.

Potassium nitrate – *See* Sodium nitrite.

Potassium nitrite – *See* Sodium nitrite.

Potassium permanganate – may be used in starch as a modifying agent. It is known to cause mild irritation and can be caustic in more highly concentrated solutions. Authorized for use in Canada and GRAS in the United States. *See also* Starch.

Potassium phosphate, dibasic (Dipotassium phosphate) – may be used in ale, beer, light beer, malt liquor, porter, stout, wine, honey wine, cider and unstandardized bakery foods as a yeast food; in cream cheese spreads, processed cheese food and spreads as a firming agent; in solid cut meat, prepared meat, solid cut poultry meat and prepared poultry meat as a sequestering agent. Commercially, it is employed as a buffering agent in the production of antifreeze solutions and as a component of instant fertilizers. Medically, potassium phosphate, dibasic is used to induce vomiting and/or diarrhea in the treatment of certain conditions. Authorized for use in Canada and GRAS in the United States. *See also* Phosphoric acid.

Potassium phosphate, monobasic (Monopotassium phosphate) – may be used in ice cream mix, ice milk, sherbet and unstandardized foods as a sequestrant and in ale, beer, light beer, malt liquor, porter, stout, wine, honey wine and cider as a yeast food. Medically, it is used to make urine more acidic. Authorized for use in Canada and GRAS in the United States. *See also* Phosphoric acid.

Potassium sorbate – may be used in apple or rhubarb jam, jellies with pectin, mincemeat, pickles, relishes, smoked or salted dried fish, tomato catsup, tomato paste, tomato purée, olive brine, margarine, unstandardized salad dressings and unstandardized foods as a Class II preservative and in bread, cheese, cheddar cheese, cream cheese, cider, wine and honey wine, as a Class III preservative. Authorized for use in Canada and GRAS in the United States. *See also* Sorbate.

Potassium stearate – is a strong alkali and may be used in chewing gum as a plasticizing agent. Authorized for use in Canada and the United States.

Potassium sulphate – may be used in ale, beer, light beer, malt liquor, porter, stout and soft drinks as a pH adjusting agent, acid-reacting and/or water-correcting agent. It can also be employed as a fertilizer and is used medically to induce vomiting and/or diarrhea. The ingestion of large amounts has been known to cause hemorrhaging in the gastrointestinal tract. Its use in Canada is governed by the Good Manufacturing Practice regulation. Authorized for use in Canada and the United States.

Potassium sulphite – *See* Sulphite.

Propane – is permitted for use in unstandardized foods as a pressure-dispensing and aerating agent. It is also used as a fuel and a refrigerant. This additive is suspected of being narcotic in high concentrations. In Canada the use of propane in foods is governed by the Good Manufacturing Practice regulation. It is GRAS in the United States.

Propene oxide – *See* Propylene oxide.

Propionic acid (Calcium propionate, Sodium propionate) – may be used in bread, cheese and some unstandardized foods as a Class III preservative. Calcium and sodium propionates are salts of propionic acids which can be found naturally in minute amounts in dairy products and certain other foods. The propionic acid and its salts used by food manufacturers are synthetically produced by chemical and pharmaceutical companies from ethylene, carbon monoxide and steam.

There have been cases of allergic disturbances, which begin in the upper gastrointestinal tract approximately four to eighteen hours after eating foods like bread and cheese (which contain propionates) and end with partial or total migraine headaches. Recent testing indicates that these compounds can suppress the conveyance of molecules into the cells, which hinders human cell growth.

Further studies on cells from human livers, intestines and chick embryos showed that antimicrobial (anti-bacteria) agents such as propionic acid were as toxic to the cells as they were to the bacteria against which they were being employed.

Researchers have concluded that a healthy human digestive system, under normal circumstances, may be able to detoxify propionates. Consumption of large amounts (due to widespread use of preservatives), however, may place too great a burden on this system and in all likelihood these agents would cause serious damage.

When these additives are used in food, they cut the losses manufacturers would otherwise incur as they inhibit the growth of mould and give stale products a fresh appearance. When sodium is added in the form of sodium propionate, it creates an additional and unnecessary burden on the body. Medically, the salts are employed by dermatologists as anti-fungal agents for athlete's foot. Authorized for use in Canada and GRAS in the United States. *See* Salt.

Propylene glycol (Propylene glycol mono- and diesters) – may be used in oil-soluble annatto and annatto margarine colour and annatto butter colour as a carrier or extraction solvent. As a point of interest, propylene glycol is also known as "antifreeze" and we question the necessity of using such a chemical in food. It is a medical fact that drinking antifreeze causes blindness, kidney failure and can even be fatal. In animal testing, large oral doses have been reported to cause central nervous system depression and kidney damage. Authorized for use in Canada and GRAS in the United States.

Propylene glycol alginate – is used in ice cream, ice milk, sherbet, cottage cheese, French dressing, mustard pickles, relishes, salad dressing, sherbet, calorie-reduced margarine, sour cream, canned asparagus, canned green beans, canned wax beans, canned peas, cream cheese, cream cheese spreads, processed cheese and spreads, cold-pack cheese foods, ale, beer, light beer, malt liquor, porter, stout as an emulsifying, gelling, stabilizing and/or thickening agent. Authorized for use in Canada and the United States. *See also* Propylene glycol and Alginate.

Propylene glycol mono- and diesters – *See* Propylene glycol.

Propylene oxide (Propene oxide) – may be used in starch as a modifying agent. Authorized for use in Canada and the United States. *See* Starch.

Propyl gallate – may be used in fats and oils, lard, shortening, dried breakfast cereals, dehydrated potato products, chewing gum, essential oils, margarine, dry flavours, citrus oils, dry cooked poultry meat, mono- and diglycerides and some unstandardized foods as a preservative. Propyl gallate is usually used in combination with one or both of BHA and BHT. In animal testing, propyl gallate has been implicated in reproductive failures, kidney and liver damage, cancer, lymphoma and allergic reactions. In 1981, the National Cancer Institute completed one of the most thorough studies on propyl gallate, which found numerous suggestions of cancer in mice and rats. According to the forty-first report of the FAO/WHO Expert Committee on Food Additives, 1993, propyl gallate decreased the hemoglobin and red blood cell count and caused morphological changes in the spleens of rats who were fed this additive during a ninety-day study. The report concluded that long-term toxicity, carcinogenicity and genotoxicity studies might be required.

Propyl gallate should be avoided as it serves no nutritive purpose in food products and alternate products should be sought wherever possible. The FDA considers it GRAS only if used in specified amounts. Authorized for use in Canada. *See also* BHA and BHT.

Propylparaben (Propyl p-hydroxybenzoate) – may be used in jams, jellies, tomato catsup, mincemeat, tomato paste, tomato purée, pickles, relishes and certain unstandardized foods as a preservative. Studies have shown that this additive suppresses the conveyance of molecules into cells, which hinders human cell growth. Further studies on cells from human livers, intestines and chick embryos showed that antimicrobial (antibacterial) agents such as propylparaben were as toxic to the cells as they were to the bacteria against which they were being employed. Researchers have concluded that a healthy human digestive system, under normal circumstances, may be able to detoxify these agents. However, consumption of large amounts (due to widespread use of preservatives) may place too great a burden on this system and in all likelihood these agents would cause serious damage. Authorized for use in Canada and the United States.

Propyl p-hydroxybenzoate – *See* Propylparaben.

Protein hydrolysate – *See* Hydrolyzed vegetable protein.

Pyrophosphate – is not considered an additive according to Division 1 of the *Food and Drugs Act*. It is used as a flavouring agent in the production of caramel and to improve the effectiveness of antioxidants in certain lards and shortenings. It could possibly cause skin and mucous membrane irritation. Pyrophosphate is considered harmless to humans although it has been proven fatal to rats in relatively low oral doses. Authorized for use in Canada and the United States. *See also* Phosphoric acid.

Quillaia (Quillaia extract) – may be used in beverage bases, beverage mixes and soft drinks as a foaming agent. It is obtained from the inner bark of a South American tree. It is presently considered to be non-toxic. Authorized for use in Canada and the United States.

Quillaia extract – *See* Quillaia.

Quinine – is a flavouring used in tonic water. It has been suggested that pregnant women avoid quinine as it may have toxic effects on the fetus. Testing has indicated that hearing may be impaired from high intakes of quinine. As a point of interest, a pilot informed us that the Regional Medical Health Examiner of Transport Canada has warned pilots that the quinine in tonic water "collects in the middle ear over a period of time which can throw off the equilibrium" and should therefore be avoided. Authorized for use in Canada and the United States.

Red #1 – *See* Artificial colour.

Red #2 (Amaranth) – *See* Artificial colour.

Red #3 (Erythrosine) – *See* Artificial colour.

Red #4 (Ponceau SX) – *See* Artificial colour.

Red #40 (Allura red) – *See* Artificial colour.

Rennet (Rennin) – may be used in cheese, cottage cheese, cream cheese, cream cheese spread, sour cream and unstandardized milk-based dessert preparations as a food enzyme. It is obtained from the aqueous extracts in the fourth stomach of calves, kids or lambs. Rennet has been used medically to aid in digestion. It has no known toxicity. Since many enzymes are now being taken from genetically altered sources, there is no guarantee that rennet will not be genetically engineered in the future. *See* Table V, page 287. Authorized for use in Canada and GRAS in the United States.

Rennin – *See* Rennet.

Rochelle salt – *See* Sodium potassium tartrate.

Saccharin – is a non-nutritive artificial sweetener derived from coal tar which is three hundred times sweeter than sugar. This now-familiar additive has had a checkered history. Saccharin was banned in 1907 by Dr. Harvey Wiley, who made it his first official act as Director of the Board of Food and Drug Inspection Agency, which later became the Food and Drug Administration (FDA). When President Theodore Roosevelt objected to the ban, a commission was appointed to review the safety of this chemical. President Roosevelt included Ira Remsen (the co-discoverer) on the commission and as might be expected, the ban was lifted.

In 1951, FDA testing found that saccharin was harmful to the kidneys of rats. In 1957, scientists were warning the FDA of the inherent dangers of saccharin, but were ignored. Saccharin remained on the GRAS list.

In 1970, the US government's concession to the accumulating evidence was to ask the National Academy of Sciences and the National Research Council (NRC) to review all the facts on saccharin. After this review, their response was that saccharin did not pose a hazard but that various tests should be conducted.

In 1970, scientists at the University of Wisconsin conducted experiments in which they implanted tiny pellets made of saccharin and cholesterol in the bladders of mice. They found that these mice had four times as many tumours as mice who had been implanted with cholesterol only. The preliminary findings in 1971 from another study in which rats were being fed saccharin indicated that it could cause them to develop bladder tumours. The results from these two studies compelled the FDA to remove saccharin from the GRAS list.

In 1974, Canadian researchers started a two-generation study on rats. It had been found that saccharin contained an impurity which might be causing the tumours – not the saccharin itself. In 1977, the results of the Canadian studies proved beyond doubt that saccharin, not the impurity, caused malignant bladder tumours. A ban on saccharin was announced by the FDA on March 9, 1977. Canadian and US officials played down the findings of these studies, emphasizing that very high doses had been used and the public was more outraged at losing saccharin than frightened by its dangers.

(Most consumers don't realize it is a standard procedure to use a high dose of a chemical in animal tests in order to increase the possiblity of detecting a carcinogen. In the Canadian tests, the use of large doses of saccharine was part of a well-established testing procedure that was necessary to determine beyond a doubt whether it was a carcinogen.)

In April 1977, due to public and industry pressure, the ban was postponed for another eighteen months in order for the NRC to review the evidence on saccharin. During that time, companies were required to print the following warning on foods containing saccharin: "Use of this product may be hazardous to your health. This product contains saccharin which has been determined to cause cancer in laboratory animals." The FDA's new commissioner, Donald Kennedy, took a firmer stand against saccharin. He said that since our bodies may already be overloaded by contaminants, we should in no way permit even a weak carcinogen to be used.

Supporters of saccharin pointed out that just because it caused cancer in rats didn't necessarily mean it would do so in humans. Since humans cannot be used as guinea pigs, it is ludicrous to argue that tests on rats are non-applicable. We must accept the results of animal testing. Furthermore, almost every substance known to cause cancer in humans has been shown to cause cancer in one or more animal species. To substantiate the fact that saccharin is a carcinogen in humans, researchers in Canada studied the medical histories of four hundred-and-eighty men who used this artificial sweetener. They found that these men were seventeen times more likely to develop cancer than people who didn't use saccharin. Researchers pointed out that it was very difficult when studying human case histories to identify a conclusive cause of cancer. For one thing, cyclamates and saccharin were frequently used together. Furthermore, humans were exposed to many other carcinogens in their food and environment.

Meanwhile, the NRC was reviewing existing saccharin studies. Researchers reported that even though saccharin is a carcinogen of low potency, they felt the fact that since it has no health benefits and would be used by large numbers of people with various health problems, its use could lead to serious health risks. They were also concerned that saccharin was found to promote the cancer-causing effects of other carcinogens. Since no one has proven how much or how little of a carcinogen it takes to cause cancer, the only solution is to refrain from using any amount of saccharin or any other proven carcinogen.

As a point of interest, although saccharin was used mainly by people attempting to control their weight, studies as far back as 1947 showed that animals that were fed saccharin (in comparison to control groups) showed increases in appetite and weight gain. This would be due to the fact that saccharin causes a decrease in blood sugar which, in turn, causes a feeling of hunger.

Saccharin was banned in France in 1950 for all uses except as a non-prescription drug. Due to the overwhelming evidence, saccharin was banned in Canada and the US in March 1977 with a provision to allow its use in certain table-top artificial sweeteners. The following excerpts from the *Food and Drugs Act* describe how it is regulated in Canada.

> E.01.004.
>
> (1) Every cyclamate sweetener that is not also a saccharin sweetener shall be labelled to state that such sweetener should be used only on the advice of a physician.
>
> (2) Commencing June 1, 1979, every saccharin sweetener shall be labelled to state that:
>
> (a) continued use of saccharin may be injurious to health; and
>
> (b) it should not be used by pregnant women except on the advice of a physician.

Saffron – may be used in jams, jellies, bread, butter, marmalade, cheeses, margarine, milk (naming the flavour), sherbet, ice cream mix, icing sugar, pickles, relishes, tomato catsup, lobster paste, smoked fish, caviar, unstandardized foods, vegetable fats and oils as a colouring agent. It is obtained from the crocus plant grown in Western Asia and Southern Europe and was previously employed in treating skin diseases. At present, it is considered non-toxic. Authorized for use in Canada and GRAS in the United States.

Salt (Sodium chloride) – is not considered an additive according to Division 1 of the *Food and Drugs Act*. It is composed of sodium and chlorine, two vital chemical substances. Although it is an essential component of every cell and of every bodily fluid, the actual amount of salt required in the diet is minimal. Due to the proliferation of processed foods, which normally contain large quantities of salt, the average person would be consuming many times more than what his or her body requires. Dr. David Reuben, author of *Everything You Always Wanted To Know About Nutrition*, aptly summed it up by saying "...the real challenge in most 'modern' countries is to keep yourself from being pickled in salt in your daily diet."

It is actually the sodium in the salt which is a major contributing factor in high blood pressure or hypertension. In the US there are approximately one hundred and fifty thousand deaths each year (approximately seventeen deaths per hour) which are attributed to high blood pressure. Excesses of salt in the early years of life will greatly increase the possibility of developing high blood pressure in later years. People who have border-line high blood pressure can control it by reducing their intake of salt. Furthermore, complications such as heart attacks, strokes, paralysis and blindness may result due to progressive and uncontrolled high blood pressure. These risks can be reduced by drastically lowering salt intake.

Some manufacturers add up to two per cent of silicon dioxide (better known as sand) to salt as an anti-caking agent. Presumably they do this as sand is even cheaper and heavier than salt, and that salt is sold by weight. Unfortunately, this is only one of many chemicals that can be added to salt. Manufacturers have a long list, some of which are aluminum calcium silicate, magnesium silicate, sodium aluminosilicate, sodium calcium aluminosilicate, tricalcium silicate, yellow prussiate of soda and sodium ferrocyanide. It is interesting to note the *Food and Drugs Act*, Division 7, allows onion, garlic and celery salt to be composed of up to seventy-five per cent salt. Authorized for use in Canada and GRAS in the United States. *See also* Silicate.

Saponin – may be used in beverage bases, beverage mixes and soft drinks as a foaming agent. It is a type of glycoside found in many plants. It is known to be poisonous to lower forms of life, and the aborigines of South Africa use it to kill fish. Saponin has a bitter flavour and when combined with water will froth when agitated. When injected into the bloodstream, it acts as a strong hemolytic which disintegrates red blood cells even when it is extremely diluted. Authorized for use in Canada.

Shellac – may be used in cake decorations and confectionery as a glazing and polishing agent as long as it does not comprise more than 0.4 per cent of the final product. It is obtained from the resinous excretion of specific insects generally found in India. During the processing, this resin can be mixed with rosin and minute amounts of arsenic trisulphide in order to achieve the desired colour. There is no arsenic used in white shellac but as the type of shellac isn't specified on the label, it is impossible to know which has been used. Researchers have proven that shellac has adverse effects on the gastrointestinal tracts of test animals. Authorized for use in Canada and the United States.

Silicate (Calcium silicate, Calcium aluminum silicate, Magnesium aluminum silicate, Magnesium silicate, Silicon dioxide, Sodium aluminum silicate, Sodium silicate, talc) – is used as an anti-caking agent in dry powdery foods such as table salt, vanilla powder, garlic powder and dry soup mixes. It is also used as a release agent and dusting agent in confections and chewing gums, and as a flavour fixative on polished white rice. Silicates are used in dry products to prevent them from becoming damp and sticking together. Just as many people put a few grains of rice in a salt shaker to absorb moisture and keep the salt free-flowing, manufacturers use silicates in many dry products for the same reason.

Silicon dioxide or silica, as it is commonly known, is simply finely pulverized rock dust. The human body absorbs only minute amounts of the silicates consumed in foods and excretes the rest in the urine and stools.

One of the silicates deserves special mention – talc or hydrous magnesium silicate. Some talcs may be contaminated with asbestos. It is well known that inhaling asbestos poses a danger to the lungs, but the hazards of ingesting asbestos fibres are not fully understood. As a note of interest, the Japanese are reported to have a high incidence of stomach cancer. In some areas of Japan, people consume large amounts of polished white rice which has been coated with California talc. California talc has been found to contain from twenty to forty per cent tremolite (calcium magnesium), a form of asbestos. Rinsing rice thoroughly not only fails to remove the talc completely but also washes away water-soluble vitamins. Just as asbestos fibres can cause lung cancer when they are inhaled, ingested asbestos fibres can cause gastrointestinal cancer. Some Canadian researchers have suggested that there is a need for more studies to determine the degree to which the walls of the digestive tract are permeated by asbestos fibres and the extent to which they are carried to other organs of the human body. Wouldn't it be safer to simply use unpolished plain brown rice and add a few grains of it to your salt shaker? Authorized for use in Canada and the United States.

Silicone – *See* Dimethylpolysiloxane.

Silicone dioxide – *See* Silicate.

Sodium acetate – may be used in starch as a modifying agent and in unstandardized foods as a pH adjusting agent, acid-reacting material and/or water-correcting agent. It is employed industrially in dyeing processes, photography and in foot warmers for its ability to retain heat. Authorized for use in Canada and the United States. *See also* Starch.

Sodium acid carbonate – *See* Sodium bicarbonate.

Sodium acid pyrophosphate – may be used in frozen fish fillets, lobster, crab, clams and shrimp to reduce processing losses and thaw drip. It is also used in processed cheese (naming the variety), cream cheese spreads and processed cheese spread as an emulsifying, gelling, stabilizing and/or thickening agent; in baking powder and unstandardized foods and in canned seafood, ice cream mix, ice milk mix, prepared meat or poultry meat as a miscellaneous food additive. Authorized for use in Canada and GRAS in the United States. *See also* Phosphoric acid.

Sodium acid sulphite – *See* Sodium bisulphite.

Sodium alginate – *See* Alginate.

Sodium aluminum phosphate – may be used in processed cheese and cream cheese spread as an emulsifying, gelling, stabilizing and/or thickening agent; in unstandardized foods as a pH adjusting agent, acid-reacting material and/or water-correcting agent. Authorized for use in Canada and GRAS in the United States. *See also* Phosphoric acid.

Sodium aluminum silicate – *See* Silicate.

Sodium aluminum sulphate – *See* Aluminum sodium sulphate.

Sodium ascorbate – may be used in flour, whole wheat flour, bread and unstandardized bakery foods as a bleaching, maturing and/or dough-conditioning agent; in ale, beer, canned mushrooms and tuna, cider, frozen fruit, head cheese, canned apple sauce, canned peaches, preserved fish, preserved meat and poultry meat and by-products, wine and unstandardized foods as a Class I preservative. It is considered to be non-toxic. Authorized for use in Canada and GRAS in the United States.

Sodium benzoate (Benzoate of soda) – may be used in jams, jellies, fruit juices (except frozen concentrated orange juice), mincemeat, pickles, relishes, tomato catsup, tomato paste, tomato purée, margarine and unstandardized foods as a Class II preservative. This chemical is generally restricted to foods with an acidic base. Medically, it is employed as an antifungal agent as well as a diagnostic aid to test liver function. Scientific testing has shown that it causes serious damage to fetuses and the FDA is studying it for further toxicity. Authorized for use in Canada and GRAS in the United States.

Sodium bicarbonate (Bicarbonate of soda, Baking soda, Sodium acid carbonate) – may be used in confectionery as an aerating agent; in salt to stabilize potassium iodide; in jams, jellies, baking powder, chocolate, ice cream mix, malted milk, infant formula, margarine and unstandardized foods as a pH adjusting agent, acid-reacting material and/or water-correcting agent and in starch as a modifying agent. Sodium bicarbonate is an alkali which is also employed medically as a gastric antacid, urinary alkalizer and as a paste for insect bites. Authorized for use in Canada and GRAS in the United States. *See also* Starch.

Sodium bisulphite (Sodium acid sulphite, Sodium hydrogen sulphite) – may be used in wine, beer, light beer, ale, porter, stout, malt liquor, cider, jams, jellies, refiners' molasses, fancy molasses, table molasses, tomato catsup, tomato paste, tomato purée, frozen sliced apples, fruit juices (except frozen concentrated orange juice), mincemeat, pickles and relishes, beverages, dried fruit and vegetables, frozen mushrooms and some unstandardized foods as a Class II preservative. Some scientists suspect that it causes genetic mutations. Authorized for use in Canada and the United States. *See also* Sulphite.

Sodium carbonate – may be used in combination with sodium hexametaphosphate for use on frozen fish fillets, frozen lobster, frozen crab, frozen clams and shrimp to reduce thaw drip; in jams, jellies, chocolate, cocoa, milk chocolate, sweet chocolate, egg white (albumen) and yolk, liquid, dried or frozen whole egg, gelatin, ice cream mix, ice milk mix, cream cheese spread, processed cheese, margarine and unstandardized food as a pH adjusting agent, acid-reacting material and/or water-correcting agent; and in starch as a modifying agent. Sodium carbonate is used as a neutralizing agent and is a strong alkali. It is commercially employed as a mouthwash, vaginal douche, water softener and a treatment for dermatitis. Authorized for use in Canada and the United States. *See also* Starch.

Sodium carboxymethylcellulose (Carboxymethylcellulose, Cellulose gum, Hydroxypropyl cellulose, Hydroxypropyl methyl cellulose, Methyl cellulose, Methyl ethyl cellulose, Microcrystalline cellulose

(MC)) – may be used in cream, French dressing, milk (naming the flavour), salad dressing, mustard pickles, relishes, cottage cheese, ice cream, sherbet, processed cheese, cream cheese and unstandardized foods as an emulsifying, gelling, stabilizing and/or thickening agent; and in non-edible sausage casings as a coating to enable peeling.

Sodium carboxymethylcellulose and its derivatives are indigestible materials that often replace good quality foods. Such substitutions can have severe detrimental nutritional effects.

Dr. Wilhelm C. Hueper, former Chief of Environmental Cancer Research at the National Cancer Institute, discovered a significant number of tumours (frequently cancerous) in rats as a result of carboxymethylcellulose. According to Dr. Hueper, sodium carboxymethylcellulose should be banned from food immediately. Furthermore, Dr. G. Jasmin of the Canadian Medical Research Council cited carboxymethylcellulose as a cancer risk as long ago as 1961.

In Canada, this additive and its derivatives are governed by Good Manufacturing Practice in the majority of cases. We fail to see the government's rationale in restricting this additive in a few foods while permitting its unlimited use in numerous other products. Authorized for use in Canada and the United States. *See also* Methyl cellulose.

Sodium carrageenan – *See* Carrageenan.

Sodium chlorite – may be used in starch as a modifying agent. It is a strong synthetic oxidizer. Sodium chlorite can be used to purify water and to bleach textiles and paper pulp. Authorized for use in Canada and the United States. *See* Starch.

Sodium citrate, tribasic (Trisodium citrate) – may be used in infant formula, apple or rhubarb jam, cottage cheese, cream, ice cream mix, jams and jellies, marmalade, sherbet, margarine and unstandardized food as a pH adjusting agent, acid-reacting material and/or water-correcting agent at levels consistent with Good Manufacturing Practice; in processed cheese, processed cheese spreads, cream cheese spreads, evaporated milk, ice cream and sherbet as an emulsifying, stabilizing, gelling and/or thickening agent; in beef blood as an anticoagulant; and in sour cream as a flavour precursor. It has been found in some cases to change urinary excretion of different drugs, causing them to be more potent or less effective. Authorized for use in Canada and GRAS in the United States.

Sodium cyclamate – *See* Cyclamate.

Sodium diacetate – may be used in bread and certain unstandardized foods as a preservative which inhibits moulds and rope-forming bacteria. It is produced by combining sodium acetate and acetic acid. Authorized for use in Canada and the United States.

Sodium dithionite – *See* Sulphite.

Sodium erythorbate – may be used in cider, frozen fruit, head cheese, meat binder for preserved meat, ale, beer, light beer, malt liquor, porter, stout, wine, preserved fish, preserved meat/poultry and dry cure employed in the curing of preserved meat, canned applesauce and unstandardized foods as a Class I preservative.

It is a non-nutritive chemical but since its composition is similar to that of vitamin C, the consumer may be led to believe that it is nutritious. In laboratory tests, biochemists found it to be rather quickly excreted. Researchers suggest that long-term studies for its mutagenic and teratogenic effects should be undertaken before it is considered safe. Sodium erythorbate has been banned in several countries yet it is still authorized for use in Canada and the United States.

Sodium furcelleran – *See* Furcelleran.

Sodium hexametaphosphate – may be used in infant formula, ice cream, ice cream mix, sherbet, mustard pickles, relishes, processed cheese and unstandardized foods as an emulsifying, gelling, stabilizing and/or thickening agent; in beef blood as an anticoagulant; in frozen fish, lobsters, crab, clams and shrimp to reduce thaw drip; in gelatin intended for marshmallow composition as a whipping agent and in canned sea foods and unstandardized foods as a sequestering agent. It is also used as a water softener and detergent. Even though phosphate is an essential nutrient, it must be kept in balance with other minerals in the diet, such as calcium. Adverse effects to the bones, kidneys and heart could result from too much phosphate. Authorized for use in Canada and the United States. *See also* Phosphoric Acid.

Sodium hydrogen sulphite – *See* Sodium bisulphite.

Sodium hydroxide (Caustic soda, Soda lye) – may be used in infant formula, ice cream mix, chocolate, cocoa, milk chocolate, sweet chocolate, gelatin, margarine and unstandardized foods as a pH adjusting agent, acid-reacting material and/or water-correcting agent and in starch as a modifying agent. Ingestion of this additive has been known to cause vomiting, prostration and total physical collapse; inhalation may result in lung damage. The FDA has restricted the use of sodium hydroxide in liquid cleaners to less than ten per cent although it is still considered GRAS for use in food. Authorized for use in Canada. *See also* Starch.

Sodium hypochlorite – may be used in starch as a modifying agent. It is also medically employed as a wound antiseptic. The ingestion of sodium hypochlorite may cause erosion of the mucous membranes and perforation of the esophagus and/or stomach. Authorized for use in Canada and the United States. *See* Starch.

Sodium lactate – may be used in margarine and unstandardized food as a pH adjusting agent, acid-reacting material and/or water-correcting agent. It is medically employed to make urine less acidic. At present, it is considered non-toxic. Authorized for use in Canada and the United States.

Sodium lauryl sulphate – may be used as a whipping agent in dried egg white/albumen, liquid egg white/albumen, frozen egg white/albumen and gelatin intended for marshmallow composition. It is also used commercially in detergents and as a component of toothpaste. This additive is capable of effecting the assimilation of certain food components such as glucose or methionine (an amino acid). Sodium lauryl sulphate is on the FDA's list for further study. Authorized for use in Canada and the United States.

Sodium metabisulphite – may be used in ale, beer, light beer, porter, stout, malt liquor, wine, honey wine, cider apple or rhubarb jams, refiners' molasses, table molasses, fancy molasses, jellies, fruit juices (except frozen concentrated orange juice), pickles and relishes, tomato catsup, tomato paste, tomato purée as a Class II preservative. Authorized for use in Canada and the United States. *See also* Sulphite.

Sodium methyl sulphate – may be used in pectin as a processing aid. At present, it is considered non-toxic. Authorized for use in Canada and the United States.

Sodium nitrate – *See* Sodium nitrite.

Sodium nitrite (Sodium nitrate, Potassium nitrate, Potassium nitrite) – is used in preserved meat to prevent botulism and to improve colour and flavour. In addition, sodium nitrate and potassium nitrate are permitted for use in certain types of cheese.

Nitrate and nitrite are very similar in chemical composition. Nitrate is considered safe until it is converted to nitrite. Other than the obvious addition of sodium or potassium to the diet, there is no great difference between the sodium and potassium forms of nitrate/nitrite.

Nitrite, in large amounts, is known to be toxic and has been reported as the cause of numerous deaths. It has the ability to change hemoglobin to methemoglobin and this is what makes nitrite so dangerous. Hemoglobin is the component of red blood cells that carries oxygen to all the cells of the body. Methemoglobin is unable to carry oxygen, therefore the conversion of hemoglobin to methemoglobin drastically impairs the blood's ability to carry oxygen. The resulting condition is called methemoglobinemia. People suffering from this condition have difficulty breathing, experience skin discolouration (bluish tinge) and can even lose consciousness which, in turn, can cause death if the levels of methemoglobin rise higher. According to FAO/WHO, babies seem especially vulnerable to methemoglobin. In addition, fetal hemoglobin is more easily changed to methemoglobin. In 1995, at its forty-fourth meeting, the FAO/WHO concluded by stating "nitrite should not be used . . . in foods for infants below the age of three months."

At one time, food manufacturers added sodium nitrite to baby food to make it look good. Once the public became informed of the risks of nitrite in baby food, they applied pressure to industry and it was voluntarily withdrawn. Nevertheless, it has been pointed out that since Canada has no regulations restricting the use of cured meats (containing nitrite) in baby foods, we have no assurance that it is not being used now, will not be used in the future or is not being imported legally now.

Studies have suggested that nitrite on its own causes cancer. However, an even more serious cause for concern is its tremendous potential for causing cancer in the form of "nitrosamines." Nitrosamines are formed when nitrites react with secondary amines which occur naturally in foods containing proteins. They are considered to be among the most powerful cancer-causing agents yet discovered. Experiments have shown that minute amounts of certain nitrosamines could cause cancer in animals. Dr. William Lijinsky, an internationally recognized authority on nitrites and cancer, told a US Senate agricultural subcommittee that there is evidence to show that nitrites in meat are the most dangerous food additives today and that they are major contributors to cancer. He went on to say that one thousand people die from cancer every day in the US alone and that most of the deaths are due to the foods people ate thirty to forty years ago. He feels that the use of nitrate and nitrite in such meats as bacon, wieners, bologna, salami, pepperoni, sausages and ham pose a significant risk to children. He verifies that nitrosamines have caused cancer in twenty-four species of animals tested at the Frederick Cancer Research Center. Other adverse reactions included miscarriages, fetal deaths and birth defects in laboratory animals.

Meat processors responded by saying that since nitrite inhibited the growth of bacteria which caused botulism, the benefits outweighed the risks. It is important to know that there are safer ways to inhibit the growth of botulism spores but these alternatives don't improve the colour or flavour of meat. The danger of botulism can be checked by proper refrigeration and cooking. Apparently the botulism spores germinate at a slow rate and refrigeration retards this growth. The meat industry concedes this point while quickly pointing out that consumers may not be as careful about refrigeration as it is. Nitrite does not destroy the spores but simply retards their germination. Dr. Ross Hume Hall describes an experiment in which bacon was held at 26°C (80°F.) The spores did not reach a toxic level (at which botulism poisoning would occur) until the tenth day. If the bacon had been refrigerated, it would have retarded the growth for an even longer period. Hall feels that consumers would agree to having meat marked with throw-away dates. We must agree especially if it would mean putting an end to sodium nitrite in our food.

The closing of small local plants and the move toward larger central plants controlled by huge corporations has created a problem for manufacturers distributing their products to consumers hundreds of miles away. They have overcome this problem with the use of sodium nitrite. It would seem that industry is not as concerned with how long consumers keep meat as they are with having extended shelf-life for their own convenience.

Some processors are now producing preserved meats free of both nitrite and botulism due to modern techniques and careful handling. Why can't this be done in all meat-processing plants? Dr. Hall recommends three alternatives to nitrite: "1. Proper sanitation in packing plants. 2. Smoking – genuine smoke, not the liquid kind that comes in bottles. 3. Salt alone will inhibit C. botulism. Salt-cured meat should be refrigerated for the short term and frozen for long term. With respect to taste, salt is the principal contributor, so absence of nitrite and nitrate would not be noticed. Note that salt-cured meats should be eaten sparingly anyway because you don't need all that salt in your diet."

Researchers have pointed out that a major portion of our nitrite consumption comes from nitrate found in drinking water and some vegetables. The use of nitrate fertilizers and the plants' natural tendency to accumulate nitrate account for the high levels of nitrate in these vegetables. Similarly, the high content of nitrate in drinking water in some areas can be attributed to the use of nitrate fertilizers that have seeped into the groundwater. Researchers have found that the addition of vitamin C and/or E to processed meats prevents, or at least retards, nitrosamine formation. It is very likely that the natural presence of these vitamins in nitrate-rich vegetables is nature's way of protecting against nitrosamines.

Canada and the US have acknowledged the danger of nitrates/nitrites by lowering the levels permitted for use in foods. The fact that nitrosamines can be formed from any amount of nitrites does not make lower levels particularly effective in preventing these formations. As we have stated previously, certain nitrosamines even in extremely tiny amounts are potent carcinogens.

Furthermore, the government has suggested that manufacturers add vitamin C and/or E to processed meats as an added protection against nitrosamine formation. We wonder how it decides the amount of these vitamins that would be needed to be effective. The proposed solution of taking a vitamin supplement daily as an added precaution does not necessarily help. Some researchers have suggested that the vitamins need to be present in the stomach at the same time as the nitrite in order to be effective.

Perhaps with increased public awareness of the dangers of these additives and less pressure from industry, the government would take more realistic action. A safer alternative would be to simply remove these dangerous chemicals and try alternate methods.

Sodium phosphate, dibasic (Disodium phosphate) – may be used in milk (naming the flavour), evaporated milk, mustard pickles, relishes, cottage cheese, sour cream, cream cheese spread and unstandardized foods as an emulsifying, gelling, stabilizing and/or thickening agent; in ale, beer, light beer, porter, stout, malt liquor, bacterial culture, cream as a pH adjusting agent, acid-reacting material and/or water-correcting agent; in ice cream mix, ice milk mix, sherbet and unstandardized foods as a sequestering agent; in frozen fish to prevent cracking of glaze and in frozen mushrooms to prevent discolouration. It can be used medically as a saline cathartic. Sodium phosphate, dibasic has been known to irritate the skin and mucous membranes. Authorized for use in Canada and GRAS in the United States. *See also* Phosphoric acid.

Sodium phosphate, monobasic (Monosodium phosphate) – may be used in processed cheese, processed cheese spread, cream cheese spread and unstandardized foods as an emulsifying, gelling, stabilizing and/or thickening agent; in ale, beer, light beer, porter, stout and malt liquor as a pH adjusting agent, acid-reacting material and/or water-correcting agent; in ice cream mix, ice milk mix, sherbet as a sequestering agent. Industrially it can be employed in boiler water treatment. It may also be used medically to make urine more acidic. Authorized for use in Canada and the United States. *See also* Phosphoric acid.

Sodium phosphate, tribasic (Trisodium phosphate) – may be used in cream cheese spreads, processed cheese, processed cheese foods and unstandardized foods as an emulsifying, gelling, stabilizing and/or thickening agent; in ale, beer, light beer, porter, stout and malt liquor as a pH adjusting agent, acid-reacting material and/or water-correcting agent. It can also be used in the manufacture of paper, in tanning leather and in detergent mixtures. Authorized for use in Canada and the United States. *See also* Phosphoric acid.

Sodium potassium tartrate (Rochelle salt, Tartrate, Cream of tartar) – may be used in processed cheese, processed cheese food, cream cheese spread and processed cheese spread as an emulsifying, gelling, stabilizing and/or thickening agent; and in apple or rhubarb jams, jellies with pectin, marmalade with pectin and margarine as a pH adjusting agent, acid-reacting material and/or water-correcting agent. It is used medically as a cathartic to induce vomiting and/or diarrhea and in the treatment of certain conditions. At present, it is considered non-toxic. Authorized for use in Canada and GRAS in the United States.

Sodium propionate – *See* Propionic acid.

Sodium pyrophosphate tetrabasic – may be used in cream cheese spreads, processed cheese, processed cheese food and unstandardized foods as an emulsifying, gelling, stabilizing and/or thickening agent. It is also permitted as a miscellaneous food additive in frozen foods such as fish fillets, minced fish, lobster, crab, clams and shrimp to reduce processing losses and to reduce thaw drip. It is employed industrially in cleansing compounds, water treatment, oil-well drilling and rust stain removal. Authorized for use in Canada and the United States. *See also* Phosphoric acid.

Sodium silicate – *See* Silicate.

Sodium sorbate – *See* Sorbate.

Sodium stearate – may be used in chewing gum as a plasticizing agent. Sodium stearate is also used in the manufacture of toothpaste and suppositories. It is used medically in the treatment of skin diseases and industrially as a waterproofing agent. At present, it is considered non-toxic. Authorized for use in Canada and the United States.

Sodium sulphate – may be used in frozen mushrooms to prevent discolouration and in unstandardized bakery foods as a yeast food. Industrially, it is used in manufacturing soaps, dyes and detergents. When taken orally, it stimulates the production of mucous in the stomach and may arrest the manufacture of pepsin, which is a natural digestive enzyme. In laboratory experiments, the adverse effects on animals fatally poisoned by sodium sulphate were confined to the intestinal tract. It is authorized for use in the United States. The Canadian food industry is bound only by the loosely defined Good Manufacturing Practice regulation.

Sodium sulphite – may be used in biscuit dough as a bleaching, maturing or dough-conditioning agent; in canned flaked tuna to prevent discolouration; in ale, beer, light beer, porter, stout, malt liquor, wine, honey wine, cider, apple or rhubarb jams, jellies with pectin, fancy molasses, table molasses, refiners' molasses, fruit juices (except frozen concentrated orange juice), gelatin, mincemeat, pickles, relishes, tomato catsup, tomato paste, tomato purée, beverages, dried fruit and vegetables, dextrose, glucose or glucose syrup, dextrose monohydrate and some unstandardized foods as a Class II preservative. Authorized for use in Canada and the United States. *See also* Sulphites.

Sodium thiosulphate – may be used in salt to stabilize potassium iodine. It is used medically to counteract cyanide poisoning, treat ringworm and mange in animals. This additive is not very well absorbed by the bowels. The FDA has listed sodium thiosulphate to be studied for its mutagenic, teratogenic, subacute and reproductive effects. Authorized for use in Canada and the United States.

Sodium trimetaphosphate – may be used in starch as a modifying agent. It is used industrially in the manufacture of detergent. Authorized for use in Canada and the United States. *See also* Phosphoric acid.

Sodium tripolyphosphate – may be used in prepared meats and unstandardized foods as a sequestering agent; in frozen minced fish, fish fillets, lobster, crabs, clams and shrimp to reduce thaw drip and processing losses; in blends of prepared fish and prepared meat as an emulsifying, gelling, stabilizing and/or thickening agent. It is also used in starch as a modifying agent. Industrially it is used to soften water and as a component of drilling fluid cleansers to curb mud viscosity in oil fields. Contact with the skin and mucous membranes can cause moderate irritation. Ingestion of sodium tripolyphosphate can cause severe vomiting and/or diarrhea. Authorized for use in Canada and the United States. *See also* Phosphoric acid.

Sorbate (Calcium sorbate, Sodium sorbate, Potassium sorbate) – may be used in apple or rhubarb jam, jellies with pectin, mincemeat, pickles, relishes, smoked or salted dried fish, tomato catsup, tomato paste, tomato purée, olive brine, margarine, unstandardized salad dressings and unstandardized foods as a Class II preservative and in bread, cheese, cheddar cheese, cream cheese, cider, wine, honey wine as a Class III preservative.

It has been suggested that when sorbates are used in combination with nitrates, they cause birth defects. Unfortunately, preventing the food industry from using sorbates and nitrite in combination is a futile exercise as we still ingest them separately in numerous other products each day. Studies of antimicrobial (antibacterial) agents performed on cells of human livers, intestines and chick embryos showed that these agents were as toxic to the cells as they were to the bacteria against which they were employed.

Researchers have determined that a healthy human digestive system, under normal circumstances, may be able to detoxify these agents. Consumption of large amounts (due to widespread use of preservatives), however, may place too great a burden on the digestive system. Authorized for use in Canada and GRAS in the United States.

Sorbic acid – may be used in apple or rhubarb jam, jellies with pectin, mincemeat, pickles, relishes, smoked or salted dried fish, tomato catsup, tomato paste, tomato purée, olive brine, margarine, unstandardized salad dressings and unstandardized foods as a Class II preservative and in bread, cheese, cheddar cheese, cream cheese, cider, wine, honey wine as a Class III preservative.

Researchers have shown that this synthetically produced chemical interrupts the function of many enzyme systems. In the human body there are numerous enzyme systems, each of which has a specific function. Therefore, any interference with them could cause problems throughout the body. Authorized for use in Canada and GRAS in the United States.

Sorbitan monostearate – may be used in imitation dry cream mix, whipped vegetable oil topping, cake mix, cake icing mix, beverage base or mix, dry soup base or mix, dried yeast and unstandardized confectionery coatings as an emulsifying, stabilizing, gelling and/or thickening agent. Various long-term studies have shown this additive to be non-toxic. Authorized for use in Canada and the United States.

Sorbitan tristearate – may be used in margarine and shortening as an emulsifying, gelling, stabilizing and/or thickening agent and in unstandardized confectionery coatings. At present, it is considered non-toxic. Authorized for use in Canada and the United States.

Sorbitol – may be used in unstandardized foods as a sweetener. Although it occurs naturally in fruit and berries, the sorbitol added to food is manufactured with chemicals. As sorbitol prevents the crystallization of sugar, food manufacturers use it as a releasing agent to maintain firmness in order to extend the food's shelf-life. Chemicals such as sorbitol, which have a low molecular weight, are more quickly broken down and absorbed by the body than those with a high molecular weight. Scientists have found this chemical can effect the absorption of vitamin B_6, which is necessary for proper functioning of blood, muscles, nerves and skin. Authorized for use in Canada and the United States.

Soyabean (Soya oil, Soya flour) – is not considered an additive according to Division 1 of the *Food and Drugs Act*. It is now grown in the midwestern United States and Canada although it was originally produced in Eastern Asia. Soyabean oil contains protein, carbohydrates, oil, ash, ascorbic acid, vitamin A and thiamine and is used in defoamers by the food industry. Debittered soyabean flour is useful in diet foods as it contains very little starch. A large percentage of soyabean crops are now genetically engineered. For more information on genetically engineered foods, please see Genetically Engineered Foods, page 109. Authorized for use in Canada and the United States.

Spices – *See* Irradiated Foods, page 127.

Stannous chloride (Tin dichloride) – may be used in processed asparagus, canned carbonated soft drinks, concentrated fruit juices (except frozen concentrated orange juice), lemon juice and lime juice as a flavour and colour stabilizer. Research indicates that as it is poorly absorbed by the body, it probably does not accumulate to any great extent but it may irritate the skin and mucous membranes. The FDA has this chemical on its list of additives to be tested for mutagenic, teratogenic, subacute and reproductive effects. Authorized for use in Canada and GRAS in the United States.

Starch (Modified starch) – may be used to give bulk, to thicken foods and to keep solids in suspension in a variety of products such as baby foods, cookies, pie fillings, sauces, gravies and pizzas. It is also used as a diluent, an anti-sticking agent, an absorbing agent, a moulding agent, a fluidizing agent, a coater, a binder, a filler and/or a stabilizer. In Canada it is not considered a food additive but it is authorized for use in foods as a nutritive material. According to the *Food and Drugs Act*, the following chemicals are permitted in or upon starch: acetic anhydride, adipic acid, aluminum sulphate, epichlorohydrin, hydrochloric acid, hydrogen peroxide, magnesium sulphate, nitric acid, octenyl succinic anhydride, peracetic acid, phosphorus oxychloride, potassium permanganate, propylene oxide, sodium acetate, sodium bicarbonate, sodium carbonate, sodium chlorite, sodium hydroxide, sodium hypochlorite, sodium trimetaphosphate, sodium tripolyphosphate, succinic anhydride and sulphuric acid. The level of use permitted for most of these chemicals is governed by the Good Manufacturing Practice regulation.

The most common sources of food starch are corn, sorghum, wheat, potatoes, tapioca and arrowroot. These raw food starches are changed with the help of one or more of the above chemicals in order to make modified starch which is more useful to food manufacturers.

Even though the original foods from which the starch is taken have nutritional value, the starch components do not. (For a more in-depth explanation of how starch is handled by the human digestive system we recommend you read *Breaking the Vicious Cycle* by Elaine Gottschall.) Even worse, modifying these starches creates a new product which has little or no nutritional merit. What you end up with is a product high in calories that is essentially devoid of vitamins, minerals, proteins, trace elements and enzymes. In her book, Elaine Gottschall states "The indigestibility of starch by even healthy people is only recently receiving attention. Some starchy foods which were assumed to be digested completely are, in fact, incompletely digested by most people. In those people with intestinal disorders, the digestibility of starch is even further effected." Add to this the possibility of chemical residues and you have what Dr. Ross Hume Hall calls a "very suspect material that should be banned." He goes on to say "I personally would avoid any product that contains modified starch."

The alkalis and acids used as starch modifiers are highly corrosive in large amounts. Other adverse effects have been identified with many of them but due to the fact that the source of the starch and the type of modification is not likely to be listed on a label, we have no way of avoiding the more dangerous chemicals or of avoiding known allergens, other than by not using products containing modified starch.

Laboratory rats have considerable difficulty digesting chemically modified starch, perhaps due to the fact that their digestive tracts have a tendency to become coated with it. Dr. Hall expresses the view that "no one can be sure how it is digested in the human digestive system but I suspect it is digested with difficulty because of its unnatural features."

In 1970, the WHO warned that considerable research needed to be conducted as modified food starches posed an unknown health risk. Modified starch is still being used in some baby foods as a thickener which reduces the need for the more nourishing and expensive ingredients that could be used in the product. Given the large number of products which contain modified food starch, it is no wonder that we have become a nation of over-fed yet undernourished people! *See also* From Lab to Larder, page 133.

Stearic acid – may be used in confectionery as a releasing agent, in chewing gum as a plasticizing agent and in foods sold in tablet form as a release agent and lubricant, where the maximum level of use is governed by the Good Manufacturing Practice regulation. This additive is a glyceride that occurs naturally in certain vegetable oils, in tallow and other animal fats and oils. It can also be produced synthetically by hydrogenating cottonseed and other vegetable oils. Stearic acid is used in suppositories, ointments and cosmetics. It is a possible irritant for people with allergies. Authorized for use in Canada and the United States.

Stearyl citrate (Isopropyl citrate) – may be used in margarine as a sequestering agent which traps the metal ions that might otherwise cause rancidity. In 1969, the FAO/WHO Expert Committee on Food Additives recommended that more complete tests be carried out with careful attention to the effects of stearyl citrate on the liver and kidneys. As antioxidants such as these are frequently not needed in food, they could be avoided by choosing brands that do not contain this additive. Authorized for use in Canada and the United States.

Stearyl monoglyceridyl citrate – may be used in shortening as an emulsion stabilizer and is produced by chemically reacting citric acid on monoglycerides of fatty acids. At present, it is considered non-toxic. Authorized for use in Canada and the United States.

Succinic anhydride – *See* Starch.

Sucralose – is permitted in or upon most table-top sweeteners, breakfast cereals, beverages, beverage concentrates, beverage mixes, dairy beverages, desserts, topping mixes, dairy desserts, frozen desserts, filling mixes, chewing gum, breath freshener products, fruit spreads, salad dressings, condiments, confections, baking mixes, bakery products, table syrups, processed fruit and vegetable products, alcoholic beverages and pudding mixes.

Sucralose was approved for use in 1991. It is one of the newest artificial sweeteners on the market and goes by the brand name Splenda™. It is produced from sugar and although it tastes, looks and performs in a similar way, it is approximately six hundred times sweeter than sugar.

Rats who were fed diets high in sucralose were found to have enlarged livers and kidneys and shrunken thymus glands. Since the main function of the liver and kidneys is to detoxify substances in the body, damage to these vital organs could have far-reaching effects. Damage to the thymus gland, which is a major part of the immune system, could lead to many health problems. Concerns have been raised about a breakdown product of sucralose that researchers feel may cause mutations. According to the Center for Science in the Public Interest (US) and the Scientific Committee for Food (Europe) more testing should be done on sucralose. Nevertheless, it is authorized for use in Canada and the United States.

Sucrose – *See* Sugar.

Sugar (Refined sugar, Sucrose, Glucose, Fructose, Dextrose, Corn syrup, Lactose, Corn sugar, Cane syrup, Invert sugar, Galactose, Maltose) – is the most commonly used food additive. Sugar is an ingredient in most cake and pastry mixes, ice cream, ice milk, frozen desserts, peanut butter, baby foods, pickles, bread, frozen dinners, salad dressing and imitation fruit drinks. The following are the sources for different forms of sugar:

- ordinary table sugar is sucrose derived from sugar cane or sugar beet
- corn syrup is derived from glucose (also known as dextrose)
- fructose is derived from fruits
- lactose is the sugar from milk
- corn sugar is from glucose
- cane syrup is from sucrose
- invert sugar is a mixture of glucose, fructose and sucrose
- galactose is derived from lactose
- maltose is produced from starch

These sugars are all refined carbohydrates, devoid of vitamins and minerals and the normal nutrients of whole foods. By any name, it is still refined sugar and too much is unhealthy for you. Since the law requires ingredients to be listed in descending order of quantity on a label (*see* B.01.008, page 151), it is not unusual to find more than one of these names interspersed throughout the list. Labelled in this way, the total amount of sugar in the product is not readily apparent. Manufacturers use sugar because it is cheap and can mask low quality ingredients and unappealing flavours.

Refined white sugar is considered one of the most harmful substances that is added to food. Refined sugar does not contain any protein, vitamins, or anything else that is necessary for the human diet. The enzyme sucrase, which is found in the intestines, breaks down sucrose into glucose and fructose. Glucose is the energizer that must be available for any work the body does. If the amount of glucose exceeds a certain level in the bloodstream, the pancreas releases a hormone known as insulin that offsets the glucose and brings the blood sugar back to normal. Glucose is stored in the liver as glycogen which can be released if the blood sugar level becomes too low.

Food manufacturers would have consumers believe the old adage that sugar gives them energy. However, if a person already feels tired and sluggish, eating a product high in refined sugar will only aggravate the symptoms. The pancreas cuts in by pouring insulin into the bloodstream to counteract this extra sugar and in a while he/she begins to feel better. Once this happens, a person feels hungry and sluggish again due to the effects of the insulin and the vicious cycle starts all over. In the meantime, the pancreas has been emptied and excess calories have been absorbed. It is on record that the Federal Trade Commission has scolded the majority of sugar promoters for claiming that sugar is good for you.

Dr. T. L. Cleave, author of *The Saccharine Disease*, is chiefly responsible for alerting the world to the dangers inherent in excess sugar consumption. According to Dr. Cleave, the process of refining sugar removes approximately ninety per cent of the sugar cane. While raw sugar provides some valuable nutrients, refined sugar provides only calories. Dr. Cleave argues that diets high in refined sugar lead to tooth decay, infections of the mouth and vagina, chronic infections of the urinary tract, excessive weight, constipation, acne, gastric and duodenal ulcers, diabetes and coronary disease. Given the disastrous effect refined sugar has on the teeth, it is not difficult to believe that it would do damage to the digestive system and to other parts of the body when it is absorbed. Dr. Cleave argues that the quality of the blood depends on the consumption of healthy, natural food. Since the heart depends on nourishment from the bloodstream, the quality of food is of major significance to the condition of the heart. Sugar is a major factor contributing to the increase of coronary disease. The problem is a tremendous one although Dr. Cleave knows the solution – avoid sugar.

The sale of sugar is a multinational business and millions of dollars are spent annually promoting sugar and sugar-laden products. Then consumers spend billions more each year in dental fees in an attempt to correct the damage done. Even worse than dental decay is the increase in degenerative diseases such as adult-onset diabetes. It should be noted that there are two forms of diabetes: juvenile and adult-onset. The juvenile form seems to be hereditary and has an absolute and life-long requirement for insulin. The adult form arises later in life and is not necessarily insulin-dependent. Diabetes is not only caused by a lack of insulin but could be due to the fact that the pancreas becomes disabled by an extended overuse of refined sugars and carbohydrates.

Insulin was discovered approximately seventy years ago and helped many people who otherwise would have died. In spite of this, there has been an increase in the death rate from diabetes. Refined sugar is used in such a wide variety of products that even the white bread, graham crackers and sponge cake that a diabetic is allowed can contain plenty of sugar. Research indicates that high-fibre diets, free from refined sugar and refined carbohydrates, can lower or eliminate the need for insulin in some diabetics. This approach to controlling diabetes should be supervised by a doctor.

Scientific studies have shown that diabetes was virtually unknown in unindustrialized societies. With the introduction of refined sugar to their diet, however, a dramatic increase in the mortality rate was noted due to diabetes, atherosclerosis and heart disease. Some of the countries studied were Iceland, Israel, South Africa, India, Trinidad and Bangladesh as well as the Inuit of Canada and Greenland. In contrast, sugar cane cutters worldwide eat sugar cane as they work and they exhibit no evidence of many of the degenerative diseases caused by refined sugars. Good alternatives are blackstrap molasses that is nearly unrefined and pure honey – but these should be kept to a minimum.

Although diabetes is known to run in families, it may not always be simply because it is hereditary. It has been suggested that if parents are big sugar eaters, then it is most likely that the children will be too. If the problem begins in infancy with heavily sugared baby formula and is followed by overly sweetened baby foods, children run a greater risk of developing diabetes than those who eat more natural foods.

Another ailment attributed to excess sugar in the diet is hypoglycemia, sometimes referred to as the "nasty personality" disease. Contrary to popular belief, hypoglycemia is not due to too little sugar in the diet but to too much. The pancreas produces more insulin than normal in order to cope with the increased blood-sugar level brought on by the consumption of sugar-laden foods. This results in a sudden drop in the blood-sugar level accompanied by symptoms similar to those of insulin shock. Some of these symptoms include irrational behaviour, fainting or blackouts, headaches, fatigue, muscle aches, numbness, irritability, depression, cold sweats, accelerated heart beat, blurred vision, nausea and loss of sexual drive. Unfortunately, this sudden drop in the blood-sugar level will trigger a desire for sweets which sets the whole cycle in motion again. Although hypoglycemia is technically the opposite of diabetes, it can be a forerunner to it due to an overtaxed pancreas which will eventually shut down. Some surveys indicate that as many as one person in ten suffers from hypoglycemia. To avoid sugar-related diseases, eat sensibly and stick to pure, unadulterated, basic foods. Authorized for use in Canada and GRAS in the United States.

Sulphite (Potassium bisulphite, Potassium metabisulphite, Sodium bisulphite, Sodium metabisulphite, Sodium sulphite, Sodium dithionite, Sulphurous acid, Potassium sulphite, Sulphur dioxide) – is used as a preservative to retard the spoilage and discolouration of many foods and beverages. In dried fruit, sulphite enhances the flavour, stabilizes the colour and prevents the formation of bacteria. In wines, sulphite will stop the growth of bacteria and yeast that could turn the wine to vinegar. Sulphite is also applied to grapes to slow their deterioration while they are being transported or stored. In freshly cut fruits and vegetables, sulphite prevents discolouration. This means that potatoes, for example, can be prepared in advance in restaurants, without turning brown.

Sulphur dioxide is a gas while the other sulphiting agents are solids. Bisulphite is the active component, which is formed when sulphiting agents are added to foods or dissolved in water. It is by combining with sugars, enzymes, oxygen and other chemicals that bisulphite protects foods from discolouration. Although little of the sulphiting agent may be left on the food, it has already been changed by the partial or total destruction of the valuable B vitamin Thiamine, possibly other members of the B complex, vitamin A and calcium. The use of sulphiting agents in food is so widespread that it could add up to a considerable daily intake and, aside from greatly lowering the nutritional value of food, could have far-reaching consequences.

People with asthma and even some people who have no history of asthma, are sensitive to sulphites. A mild reaction may include hives, weakness and difficulty in breathing; a moderate reaction may include dizziness, abdominal pain, diarrhea and vomiting and a severe reaction can cause loss of consciousness and even death. In fact, the FDA has reported seventeen deaths that have been associated with sulphites.

There are other good reasons for avoiding sulphites even if you don't suffer from asthma. Sulphites have caused tumours in rats and genetic mutations in micro-organisms. Heavy emissions of sulphur dioxide are responsible for the devastation of considerable areas of vegetation in mining cities such as Sudbury, Ontario. Authorized for use in Canada and the United States.

Sulphur dioxide – *See* Sulphite.

Sulphuric acid – may be used on coffee beans to improve the extraction yield of coffee solids. It may also be used in ale, beer, light beer, malt liquor, porter, stout as a pH-adjusting agent, acid-reacting material and/or water-correcting agent and in starch as a modifying agent. The maximum level of use for sulphuric acid in the majority of these foods is governed by the Good Manufacturing Practice regulation, which leaves the level of use up to the discretion of the manufacturer. Authorized for use in Canada and the United States.

Sulphurous acid – *See* Sulphite.

Sunset yellow FCF (Yellow #6) – *See* Artificial colour.

Talc – *See* Silicate.

Tallow flakes (Beef tallow) – are not considered an additive according to Division 1 of the *Food and Drugs Act.* They are obtained from the fat of sheep and cattle in North America and are used as a defoaming agent in the production of certain food and nonfood products. Testing of miniature pigs over a one-year period produced moderate-to-severe hardening of the arteries much the same as is found in humans. Authorized for use in Canada and GRAS in the United States.

Tamarind – is not considered an additive according to Division 1 of the *Food and Drugs Act.* It is obtained from dried tamarind fruit grown in Africa and the East Indies and is used as a natural food flavouring. It can also be used as a laxative. It is not considered toxic. Authorized for use in Canada and the United States.

Tannic acid (Tannin) – is used in cider, honey wine and wine as an emulsifying, gelling, stabilizing and/or thickening agent and in chewing gum to reduce adhesion. It is produced by a combination of chemicals obtained from the bark and other parts of numerous trees, shrubs and plants. It is also found in tea and coffee. Scientists suspect tannic acid of being a weak carcinogen as it has caused tumours and liver damage in test animals. In Canada, the use of this additive is restricted to certain percentages in cider and wine, however the level in chewing gum falls under the Good Manufacturing Practice regulation. GRAS in the United States.

Tannin – *See* Tannic acid.

Tartaric acid – may be used in beer, light beer, ale, porter, stout, honey wine, malt liquor, cider, jams, jellies, baking powder, salad dressing, French dressing, ice cream mix, mayonnaise, canned pears, canned strawberries, processed cheese, margarine and unstandardized foods as a pH adjusting agent, acid-reacting material and/or water-correcting agent; in fats, oils, lards, shortening, mono- and diglycerides and certain unstandardized foods as a Class IV preservative. It occurs naturally in fruit and can also be commercially produced from by-products of wine production. Lifetime feeding tests have so far shown this additive to be harmless in the quantities presently used. Authorized for use in Canada and GRAS in the United States.

Tartrate – *See* Sodium potassium tartrate.

Tartrazine (Yellow #5) – *See* Artificial colour.

Textured vegetable protein (Isolated soy protein, TVP) – is not considered an additive according to Division 1 of the *Food and Drugs Act*. Textured vegetable protien can be produced by chemically modifying soy protein flour in various ways. For example, the flour can be dissolved in lye and then treated in an acid bath so that the protein is deposited in finely spun threads which are then wound on a spool. It is then glued, coloured, flavoured and cut into meat-sized chunks. Food manufacturers like its low cost and versatility as it can be made to taste like hamburger, sausage, chicken, pork, ham, nuts and nearly anything else while remaining inexpensive to produce. It can be used in conjunction with meats to increase bulk or to replace a portion of the original meat. Textured vegetable protien is often used in restaurants, schools, hospitals and most of all, fast food outlets. Isolated soy protein is merely the untreated protein extracted from soyabeans. Authorized for use in Canada and the United States.

Tin dichloride – *See* Stannous chloride.

Titanium dioxide – may be used in bread, butter, concentrated fruit juice (unfrozen varieties), ice cream mix, icing sugar, jam and jellies with pectin, liqueurs, milk, pickles, relishes, sherbet, smoked fish, lobster paste, caviar, tomato catsup, liquid, dried or frozen whole egg, vegetable fats and oils, margarine, a variety of cheeses and unstandardized foods as a colouring agent. It is found naturally in certain minerals and can be produced commercially in various ways. Lung damage can occur after inhaling large quantities of the dust. It has been suggested that its effects on the human body are unclear and more research needs to be done. The European Community countries have banned its use. Authorized for use in Canada and the United States. *See also* Artificial colour.

Tragacanth gum – may be used in salad dressing, French dressing, mustard pickles, relishes, processed and cream cheeses, cottage cheese, ice cream, sherbet, calorie-reduced margarine and unstandardized foods as an emulsifying, gelling, thickening or stabilizing agent. It is obtained from a small bush in Iran and because of its high resistance to acids, it is ideal for use in acidic foods. Tragacanth gum has been re-evaluated in the US and is considered GRAS in certain percentages. Nevertheless, when taken in large amounts, it can cause allergic reactions, diarrhea, gas and constipation. Authorized for use in Canada and GRAS in the United States.

Triacetin (Glyceryl triacetate) – may be used in cake mixes as a wetting agent; in flavours (naming the flavour) and in unstandardized flavouring preparations as a carrier or extraction solvent. It is produced by adding acetate to glycerine. Commercially it is employed as a fixative in perfumes and industrially it is used as a solvent in the manufacture of celluloid and photographic films. It can be used medically as a topical antifungal agent. Authorized for use in Canada and the United States.

Tributyrin (Glyceryl tributyrate) – may be used in flavours (naming the flavour) and unstandardized flavouring preparations as a carrier or extraction solvent. At present, it is not considered toxic. Authorized for use in Canada and the United States.

Tricalcium phosphate – *See* Calcium phosphate, tribasic.

Triethyl citrate – may be used in liquid egg white/albumen, frozen egg white/albumen as a whipping agent and in certain flavours and unstandardized flavouring preparations as a carrier or extraction solvent. It is used as a plasticizer in nail polish. Authorized for use in Canada and GRAS in the United States.

Trisodium citrate – *See* Sodium Citrate, tribasic.

Trisodium phosphate – *See* Sodium Phosphate, tribasic.

Turmeric – may be used in bread, butter, concentrated fruit juice (unfrozen varieties), ice cream mix, icing sugar, jam and jellies with pectin, liqueurs, milk, pickles, relishes, sherbet, smoked fish, lobster paste, caviar, tomato catsup, liquid, dried or frozen whole egg, vegetable fats and oils, margarine, a variety of cheeses and unstandardized foods as a colouring agent. It is obtained from a herb found in the East Indies that has a bitter taste. It is not considered toxic. Authorized for use in Canada and GRAS in the United States.

TVP – *See* Textured vegetable protein.

Urea – the use of urea in honey wine and wine as a yeast food was revoked on December 18, 1986. It is found in certain bodily fluids such as urine and is the product of protein metabolism. Urea is synthetically manufactured by combining ammonia and carbon dioxide under pressure to produce ammonium carbamate which is then dehydrated to form urea and water. It is also widely employed in fertilizers and animal feeds (due to its high nitrogen content). Urea is commercially used in the manufacture of plastics, skin creams and ammoniated dentrifices and medically, in the reduction of body water and intracranial and eye pressure. Authorized for use in Canada and GRAS in the United States.

US red dye #4 (Red Dye #4) – *See* Artificial colour.

US yellow dye #5 (Yellow Dye #5) – *See* Artificial colour.

US yellow dye #6 (Yellow Dye #6) – *See* Artificial colour.

Vanilla – *See* Vanillin.

Vanillin (Ethyl vanillin, Vanilla) – is not considered an additive according to Division 1 of the *Food and Drugs Act*. It is a synthetic chemical used in chocolate bars, ice cream, soda pop, bakery goods, candy and literally hundreds of other products as a flavouring. It is produced from eugenol or from the by-products of the pulp industry. Dr. David Reuben states, "If you feed a mouse 1/20 of an ounce of vanillin for each pound of his body weight, you will have a very dead mouse – and very quick. That's about 2.5 ounces of vanillin for a fifty pound child." Apparently food manufacturers use vanillin in place of natural vanilla because it is cheaper. Food manufacturers justify this practice by pointing out that vanilla production at present does not meet the demand. We must agree with Dr. Reuben's solution to this problem, "Plant more vanilla beans." Authorized for use in Canada and GRAS in the United States.

Vegeteable lutein – *See* Xanthophyll.

Violet #1 – *See* Artificial colour.

Vitamin C – *See* Ascorbic acid.

Whey – is not considered an additive according to Division 1 of the *Food and Drugs Act*. It is obtained as a by-product from milk. It is not considered to be toxic. Authorized for use in Canada and the United States.

Xanthan gum – is a commercially produced gum which may be used in French dressing, salad dressing, calorie-reduced margarine, processed cheese, cream cheese, cottage cheese, mustard pickles, relishes, ice cream mix, ice milk mix, sherbet, cream for whipping and unstandardized foods as an emulsifying, stabilizing and/or thickening agent. It is not considered toxic. Authorized for use in Canada and the United States.

Xanthophyll (Vegetable lutein) – may be used in bread, butter, concentrated fruit juice (unfrozen varieties), ice cream mix, icing sugar, jam and jellies with pectin, liqueurs, milk, pickles, relishes, sherbet, smoked fish, lobster paste, caviar, tomato catsup, liquid, dried or frozen whole egg, vegetable fats and oils, margarine, a variety of cheeses and unstandardized foods as a colouring agent. It was first isolated from egg yolk but, perhaps due to its high cost, it is now isolated from flower petals. Xanthophyll is not converted into vitamin A as most carotenoids normally are. Authorized for use in Canada and provisionally permitted for use in the United States. *See also* Artificial colour.

Xylitol – is a rare food sugar used in unstandardized foods as a sweetener. It is manufactured from waste products from the pulp industry. Chemicals such as xylitol that have a low molecular weight are more quickly broken down and absorbed by the body than those with a high molecular weight. This means that the chemical is more apt to be stored in the body where it would have a greater potential to cause harm. Conversely, a chemical with a higher molecular weight would probably be excreted more quickly and be relatively unabsorbed by the body. When given intravenously, it has been linked to numerous medical disorders associated with disturbed metabolism in humans. Other adverse reactions in humans include acidosis, loss of body fluids, kidney stones, kidney failure, nausea, disorientation, unconsciousness and death. The preliminary reports of the FDA specify xylitol as a cancer-causing agent. Despite the apparent human toxicity, the Canadian government regulates its use under the Good Manufacturing Practice regulation. Authorized for use in the United States.

Yellow #1 – *See* Artificial colour.

Yellow #2 – *See* Artificial colour.

Yellow #3 – *See* Artificial colour.

Yellow #4 – *See* Artificial colour.

Yellow #5 (Tartrazine) – *See* Artificial colour.

Yellow #6 (Sunset Yellow FCF) – *See* Artificial colour.

2

Genetically Engineered Foods

We live in an era where innovative technology pervades every aspect of our lives, including the genetic makeup of our food. The changes in the genetic makeup of the plants and animals grown for food was traditionally accomplished by selective breeding and cross-breeding within the same species. In the last two decades, however, science has begun to tamper with and reshuffle genetic material between plants and animals. These developments are only recently being brought to light and we owe it to ourselves and future generations to make the effort to understand the implications of this technology.

Deoxyribonucleic acid (DNA) is the substance that carries the hereditary genetic information of living matter. *Genes* are segments of DNA that provide the blueprint for every part of a person, plant, bacteria, virus, fungus, in short, for every living organism. Genes determine the characteristics or traits that make each organism unique such as colour, height, nutritional quality or even tolerance to frost. The genetic material that makes each organism unique is found only in those specific organisms. Thus, you would not normally find the genes of a plant in a cow or vice versa. *Genetic engineering* is the technology that allows scientists to transfer hereditary information (in the form of genes) directly from one organism (a donor) to another organism (a recipient). If the transfer is carried out on organisms from different species, for example from a cow to a plant or from a fish to a tomato, the resulting genetically engineered organism is called transgenic. The genetic makeup of the transgenic organism is different from that of the donor or the recipient. A new living organism has been created with the ability to grow, reproduce and mutate.

Genetic engineering has been adapted to a wide variety of agricultural uses, among others. For instance, let's look at a farmer who has a problem with weeds in his tomato field. He would plant tomato seeds or plants that had been genetically engineered to be herbicide-resistant. Consequently, when the farmer sprays his fields with a weed killer, it might destroy everything in sight but shouldn't hurt his tomato plants.

The herbicide resistant gene inserted into the tomato makes the plant tolerate greater amounts of weed killer.

Another farmer might want to transport his tomatoes to a market thousands of miles away. He would be interested in growing a tomato whose genes had been manipulated to prevent the cell walls from collapsing even if the tomato was ripe. This would enable the tomatoes to survive the trip. Another farmer who faced problems with pests would want a tomato plant that discouraged insects, so his choice would be a tomato into which a pesticide gene had been inserted. When the insects began to eat his tomato plants, they would either perish after a few mouthfuls or would decide to lunch elsewhere.

Here is how this technology works. All organisms have built-in defense mechanisms which normally screen out foreign genetic material. These species barriers are provided by nature to prevent cross-mixing of species genes and the occurrence of strange mutations. Scientists found they could overcome the species barrier in a recipient by using *vectors*. Vectors are vehicles which transfer or smuggle a gene from one species into a cell of another species and eventually into all the new cells generated by the recipient. The most successful vectors are bacteria or viruses because they have the ability to invade the genetic structure of the host's cells. Once the transfer was made, the scientists needed a quick method to confirm that the gene had been incorporated. They found they could insert an antibiotic-resistant gene, as a *marker*, at the same time they inserted the foreign gene and the vector into the host. A marker is a gene which will not be harmed by antibiotics but will continue to grow and thrive. Now, by flooding a petri dish (containing a specimen of the host organism) with an antibiotic, they could quickly tell if the gene transfer was successful. If the host organism was unharmed by the antibiotic, then they had made a successful gene transfer. Conversely, if the host organism was destroyed, the transfer was a failure as the antibiotic resistance from the marker gene had not been picked up. When gene transfer is successful, the new transgenic organism could contain not one, but three, new genes.

1. a pesticide gene
2. a bacteria or virus gene (the vector or vehicle)
3. an antibiotic resistant gene (as a marker)

The following chart shows some of the crops that have been manipulated by inserting genetic material from foreign species.

Crop	Source of New Genes	Purpose
tomato	flounder	reduced freezing damage
	virus	increased disease resistance
	bacteria	reduced insect damage
potato	chicken	increased disease resistance
	virus	increased disease resistance
	bacteria	herbicide tolerance
	giant silk moth	increased disease resistance
	greater waxmoth	reduced bruising damage
corn	bacteria	herbicide tolerance
	firefly	introduction of marker genes
	wheat	reduced insect damage
melon, squash and cucumber	virus	increased disease resistance
rice	bacteria	reduced insect damage
	bean and pea	new storage proteins
lettuce and cucumber	tobacco petunia	increased disease resistance

One of the most alarming features of this technology is that gene transfer between species may cause the spread of dangerous diseases across species barriers. The claim that the vectors or viruses used to insert genes are inactivated or crippled strains (meaning they are no longer capable of causing disease) is refuted by opponents of genetic engineering who warn that inactivated viruses have the capability of becoming re-activated. They say it has been proven that DNA of such viruses and bacteria, when incorporated into food, is capable of surviving passage through the digestive system from where it can find its way into the bloodstream and hence, into any of the cells in the body. The viruses can then break down cell walls of organisms already infected by other viruses, genetically combine forces and begin to grow again. Through recombination, these viruses and bacteria can generate new and more virulent viruses and bacteria capable of causing diseases and cancers in more than one species, making them almost impossible to eradicate.

Scientists have linked the emergence of pathogenic (disease-causing) bacteria, antibiotic resistance and more virulent viruses to horizontal gene transfer.

Horizontal gene transfer is the transfer of genes to unrelated species, by infection from viruses, through pieces of DNA taken up by cells from the environment, or by unusual mating taking place between unrelated species. Genetic engineering is specifically designed to break down species barriers in order to transfer genes between species that do not normally interbreed.

They feel that genetic engineering has increased the variety of hosts for rare and potentially fatal viruses such as monkeypox and hantavirus. Previously these viruses were contracted from contact with infected rodents. Now, however, they are reported as being transmitted directly from human to human. The increase in recent years of debilitating conditions such as fibromyalgia and chronic fatigue syndrome is also thought to be linked to genetic engineering. Dr. Mae-Wan Ho, Professor of Biology Department, Walton Hall University, warns that "horizontal gene transfer is responsible for the creation of new pathogens. Totally unrelated pathogens are now showing up with identical virulence and antibiotic-resistance genes." According to Dr. Ho, "The escalation in the emergence of pathogens and antibiotic resistance over the past decade coincides with the commercialization of genetic engineering biotechnology."

In fact in 1996, the World Health Organization (WHO) reported the emergence of at least thirty new diseases over the past twenty years including AIDS, Hepatitis C and Ebola. At the same time, infectious diseases such as tuberculosis, diphtheria, cholera and malaria, which were once considered under control, are now making a comeback. Outbreaks of infections such as E.coli, meningitis and streptococcus are occurring more frequently. Furthermore, these new strains seem to be more virulent than their predecessors and are becoming increasingly difficult to treat due to their resistance to almost all known antibiotics. For example, a strain of the bacteria staphylococcus aureus isolated by scientists in Japan is resistant to all antibiotics including vancomycin, which is considered the "last resort antibiotic". With international travel so commonplace, one can see how easily a disease can become global almost overnight.

The British physicist, Dr. Joseph Rotblat, winner of the 1995 Nobel Prize for his campaign against nuclear weapons compared the potential for devastation from genetic engineering to that of nuclear weapons. He stated: "My worry is that other advances in science may result in other means of mass destruction. . . . Genetic engineering is quite a possible area, because of these dreadful developments that are taking place there." Dr. Rotblat's comments should be sufficient to compel us to examine this new technology and the possible global impact of releasing genetically altered organisms.

Increased meddling in food production has also resulted in more cases of poisonous and unhealthy food. For example, the end result of Bovine spongiform encephalopathy (BSE), also known as "mad cow disease," remains to be seen as it is still too early to predict how many other people will die from it. Researchers feel that it resulted from feeding cattle totally unnatural food. In March, 1996, British officials announced that the deaths of ten people were linked to eating infected beef and British scientists publicly admitted that BSE, a fatal disease, could be transmitted from one living creature to another. Since 1986, thousands of British cattle have died of mad cow disease, which riddled the ailing cow's brain with tiny holes that left the brain looking like a sponge. Experts believe the brain-wasting sickness was introduced into the cattle by feeding cows a protein supplement, in the form of meal, that was being produced with the ground-up remains of dead cattle. Other ruminants such as sheeps' carcasses, which were infected with a similar illness called "scrapie," were found in the meal as well. This scandal virtually brought the British meat industry to a standstill as Europe and other countries banned imported British beef. Although Britain had banned feeding dead sheep parts to British cattle in 1988, when the statistics were reviewed in 1995 over half of all the British cattle that had developed BSE were born after the ban was implemented. Mad cow disease exposed fundamental flaws in food safety regulations and reminded people that science is never infallible. There is an analogy to be made here. Genetic engineering has introduced unnatural foods into the human food supply. As in mad cow disease, there is also the possibility of passing trans-species diseases to humans by inserting genes from animals, bacteria and viruses into our foods.

Along with the increase in disease, allergies have also become more common. We probably all know somebody who has food sensitivities. Many allergies are more than simply an inconvenience, some are deadly. The insertion of foreign genes into crops has the potential to create allergens in foods that previously did not trigger allergic reactions. Since genetically engineered foods do not require specific identification on labels, known allergens would be even more difficult to avoid. For instance, a person with a known allergy to fish could have a reaction to a tomato which had a fish gene incorporated without being aware of the source of the allergen. That's good news for the company who produced the tomato as the allergen would be almost impossible to trace, but not good for the person with the allergy. This already happened with soyabeans that were genetically engineered with genes from the Brazil nut. The allergen in the Brazil nut was transmitted to the soyabean and caused allergic reactions in sensitive people. This is disquieting especially when one considers the soyabean is normally allergy free and therefore safe for most people to eat. Fortunately, the allergen in this case was detected and production of this soyabean was discontinued. However, allergic reactions to all combinations and possible mutations may not be so easy to detect, much less predict.

A dramatic case of genetic engineering going awry was the tryptophan disaster. In 1987, the Showa Denko Company of Japan began using genetic engineering to reduce costs and speed up the production of the food supplement tryptophan. Genetically engineered bacteria was used in the commercial production of this tryptophan which was sold on the American market in 1988. According to US law, the company was allowed to sell the genetically engineered supplement without testing it for safety as they had been selling non-genetically engineered tryptophan for years without problems. Therefore, the new product was considered "substantially equivalent," or safe.

The toxin in the genetically engineered tryptophan caused a condition called Eosinophilia myalgia syndrome (EMS). It was characterized by symptoms such as swelling and cracking of the skin, inflammation of the joints, headaches, fatigue, suppression of the immune system, heart problems and crippling muscle pain. The toxin, a by-product of the introduced gene, is said to have constituted less than 0.1 per cent of the finished product, yet it was enough to kill thirty-seven people and leave another 1,500 permanently disabled. It took months to track the toxin back to the genetically engineered bacteria which Showa Denko had used. This delay was caused in part by the fact that the tryptophan had not been labelled "genetically engineered" and therefore was hard to distinguish from the traditionally prepared supplement.

Dr. John B. Fagan, an American biologist and a spokesman for Consumers' International commented that genetic engineering was implicated in the tryptophan scandal because the original, non-genetically engineered bacteria had never been shown to contain the toxin. In addition, other manufacturers of tryptophan who used non-genetically engineered bacteria never had outbreaks of Eosinophilia myalgia syndrome. Dr. Fagan concluded that ". . . it is highly likely that genetic engineering was the determining factor in generating this toxin." What makes genetic engineering particularly dangerous is that no one can predict the new toxins and allergens that will develop as a result of recombination. Since the genetic code is enormously complex, no one knows where the inserted gene will land or if the new gene will act differently than expected when placed in a new host. These are a few areas that must be given serious consideration.

Researchers predict irreversible changes in the delicate balance of nature as a result of cultivating genetically engineered food plants. Existing genetic plant structures have evolved over millions of years into an infinitely complicated interdependent eco-system. Scientists are altering nature's balance with changes which could not happen naturally. This is being done too quickly, with little, if any, long-term testing. Genes in plants can and will replicate, spread and recombine into new species. Once released, they can never be recalled. Therefore, planting large areas of agricultural land with genetically engineered seeds will have far-reaching agricultural consequences.

One hazard of genetically altered plants which has received little consideration is their effect on soil insects, burrowing animals, earthworms, seed and insect-eating birds and non-target mammals. We have already begun to see examples of environmental damage to soil micro-organisms and insects as a result of gene-altered organisms. In the early 1990s, the unplanned escape of a genetically altered organism, klebsiella bacteria, infected some wheat fields in the United States. It destroyed the nutritive bacteria of the fields, rendering the fields sterile and capable of growing wheat only a few inches high.

In another instance, studies at the Scottish Crop Research Institute (SCRI) showed that a pest-resistant gene which was inserted into potatoes to suppress the growth and reproduction of aphids damaged beneficial insects as well. Ladybugs normally eat aphids and are considered a beneficial insect. In order to discover what effect, if any, the introduction of the gene-altered potato would have on beneficial insects, the researchers fed ladybugs aphids that had eaten these potato plants. They found that the female ladybugs lived only half as long as they normally would and only one third of their fertile eggs hatched. These results invalidated the theory that the toxin would not cause harm beyond the target pest. The SCRI recommended that all crops genetically engineered to be pest-resistant should be studied in this manner in order to avoid upsetting the fragile balance between beneficial insects and pests in the environment.

An example of a pesticide gene which is being inserted directly into certain food plants is the soil bacteria, *Bacillus thuringiensis* (Bt). This bacteria naturally produces a pesticide which has been used externally on plants for approximately thirty years. Once this gene is inserted in the recipient, it will continue to produce the toxin throughout the growing season in order to destroy the predator's digestive tract. Researchers claim the toxin will be confined to the above ground growth and will not enter the edible portions.

Home gardeners will be surprised to learn that they may be growing genetically altered seeds unknowingly. Genetically engineered potatoes (Newleaf™) are offered for sale in some seed catalogues. The advertisements claim the plant will be protected from the potato beetle because it contains a gene from the *Bacillus thuringiensis* var. *tenebrionis* (Btt). Apparently these plants will be completely protected for the entire growing season without the use of more insecticides. These potatoes are supposed to be comparable to non-genetically engineered varieties in size, shape, texture and taste. The advertisement asserts these potatoes ". . . have absolutely no effect on other insects, animals, people or the environment."

Bio-engineering is not that exact! Nobody can give that assurance. If Bt is toxic to the potato beetle, how can we be certain it won't also have human toxicity? Some researchers say that even if the toxin did enter the edible portions of the plants, human stomach acid would destroy it. Can scientists guarantee that it will not harm human stomachs in the same manner as it harms the insects? What testing has been done to this end? Are the effects of this toxin multiplied if a person has an intestinal disorder or low stomach acid? Surely, before such products are marketed, these questions should be answered.

There have even been reports that the genetic insertion of Bt may not be completely effective in protecting some plants against insect assault. This would force farmers to use even more pesticides – rather than less, as these companies are promising. As a point of interest, in 1996, a group of Texas cotton farmers brought a class action suit against Monsanto. Monsanto had promised that their engineered cotton, Bollgard™, which contained Bt, would destroy the bollworm caterpillar without the use of additional chemicals. It did not, however, protect the young shoots against the caterpillar. The result was the destruction of up to sixty per cent of plants in some fields with approximately one billion dollars in damage for that harvest period.

Many scientists are concerned that genetically engineered crops which escape their boundaries or fields will become dangerous weeds in nonagricultural settings. Weeds can be defined as plants that happen to be in the wrong place at the wrong time. A plant that you might consider an unwanted weed, like grass in your garden, is a desirable plant when it grows on your lawn. A broader definition would be any plant that is not intentionally sown and whose undesirable effects or characteristics far surpass its good points.

While agricultural plants are prone to escape and establish themselves in other areas, the addition of new genetic plant material increases the rate of flow from one population to another via pollen. For instance, in one field test seventy-two per cent of non-genetically engineered potatoes had picked up the novel gene from a genetically engineered potato crop that was planted a kilometre away. Experiments with crops grown more than one kilometre apart showed that gene transfer is common over such great distances. In fact, thirty-five per cent of these crops picked up the new gene.

When a gene is taken from one species and inserted into another, it is called a *novel gene*. Novel gene out-flow to wild species will produce superior plants that will then re-invade and displace other crops. Should this occur, these new gene-altered weeds will reduce the crop yield by competing for nutrients, water, light or space and, if harvested with crops such as wheat, will reduce the quality of the harvest.

Novel genes could also enhance the weediness of an existing pesky weed, turning it into a major nuisance. For instance, herbicide-tolerant genes from genetically engineered crops could outflow to weeds, thus making the weeds resistant to the herbicides applied to them. This would make the weeds almost impossible to eradicate. Farmers would be faced with the choice of having either to find a new herbicide that is effective against the new weed or multiply the quantity of herbicides used. Either way, the farmer risks effecting his crops.

Genetically engineered plants whose “designer genes” are known to migrate vast distances could cause even greater destruction than the following plants, which were non-indigenous to the areas into which they were introduced. They subsequently flourished and caused massive environmental damage.

Kudzu was brought into the States for ornamental purposes, erosion control and as forage for cattle. It has invaded forests and field borders and is now out of control. This tenacious weed infests over seven million acres in the Southern US and repeated attempts to remove it have failed.

Purple Loosestrife was introduced from Europe as an ornamental plant. It has aggressively invaded lake shores and wetlands, replacing native aquatic vegetation and disrupting the eco-systems by causing changes in the animal food chain. It has invaded the flood plains of the St. Lawrence River and is a serious threat to marsh plants throughout the north central regions of the United States.

Hydrilla was brought in from South America for use in the aquarium trade. Florida spends seven million dollars annually combating its choking of a number of lakes, reservoirs and waterways.

Melaleuca was introduced as an ornamental tree. It has invaded wetlands, infesting 450 thousand acres in the United States. Melaleuca is replacing cypress forests, sawgrass marshes and other habitats at the rate of fifty acres a day.

Commercial production of plants which are gene-altered to be resistant to certain pesticides and herbicides would greatly increase the use of such chemicals, leading to further contamination of soil and water. In fact, these plants are already being produced. Monsanto, for example, has created Roundup-Ready™ seeds which will grow into crops that are tolerant to Mosanto's herbicide, Roundup™. Although Monsanto claimed that the half-life of Roundup residue was no more than sixty days in water and soil, crops that were planted a year after the last spraying were found to contain such residues.

Danish studies show that Roundup subsists in the soil for as long as three years and can be absorbed by subsequent crops. Residues of glyphosate (the active ingredient in Roundup) have been found in strawberries, lettuce, carrots, barley and fish. Some lettuce, carrots and barley crops which were harvested a year after the application of Roundup contained glyphosate residues.

German studies have shown that glyphosate increases levels of plant estrogen (female hormone) when sprayed on crops. These increases in estrogen have produced unpredictable adverse effects on wildlife, even causing male fish to lay eggs! Other hazards of Roundup are

- illnesses among agricultural workers
- damage to the reproductive systems in mammals
- injury to beneficial soil micro-organisms, earthworms and beneficial insects
- decreased wild plant diversity
- pollution of water tables

In addition to seeds which are resistant to specific herbicides or pesticides, companies are now creating *terminator seeds*. Scientists have discovered a way to insert blocks of genes into crop plants that renders the crops incapable of reproducing. Terminator seeds will develop into mature plants but they will not produce seeds that farmers can save for subsequent plantings. In order for the manufacturer to produce more seeds to sell farmers the following year, they spray the plants with a patented chemical that shuts off the blocker genes and allows their terminator crop to produce seeds that will germinate.

Another concern of terminator seeds is that companies can patent and keep the chemical spray formula a trade secret. Furthermore, the residues from the chemical spray will undoubtedly find their way into food and if the formula is a trade secret, it will be next to impossible to assure public safety from cancer, birth defects, allergies, autoimmune disease or any of the many other adverse effects that chemical sprays have been known to cause in the past. When bureaucrats start allowing such formulas to be considered trade secrets, none of us are safe!

Companies such as Monsanto have been consolidating their power through mergers and acquisition of seed companies and smaller farms as well as by producing genetically altered organisms that would be resistant to only their own herbicides and pesticides and then patenting them. It is now conceivable that a few huge companies could end up controlling the food supply from the seed to the end product, which would force farmers to purchase all their seeds and herbicides from the gene merchants.

Currently only a handful of giant corporations are controlling agricultural biotechnology and they are in the process of further mergers. For instance, *The Wall Street Journal* recently reported that Dow Chemical is forming a new company called Advanced AgriTraits. This company ". . . will act as a clearinghouse where other companies can go to develop, license and market new genes for improved crops." The article went on to say that a few chemical giants ". . . have invested billions of dollars to boost their biotechnology research efforts and lock up key market positions in life sciences." It would appear that corporate control of the food supply is the way of the future. . . .

Organic Foods vs Genetically Engineered Foods: Issues and Debate

Currently, only "certified organic" foods and seeds are guaranteed not to be genetically engineered. The manufacturers of genetically engineered products, however, are appealing to governments throughout the world to legislate new laws that will allow them to define genetically engineered products as organic because real organic food has a valuable clean image whereas genetically engineered foods are considered sullied. If manufacturers are able to blur the distinction between organic and gene-altered foods, they would eventually control the entire food chain. Many legitimate organic farmers would be forced out of business as they could not compete with cheaper genetically engineered food products.

Indeed, manufacturers of genetically engineered foods, irradiated foods and crops fertilized with sewage sludge could take advantage of the growing market for foods labelled "organic." This move to redefine the term organic caused tremendous furore in the US in early 1998 when manufacturers lobbied the government to relax the standards for organically grown foods. Fortunately, approximately 150,000 angry citizens opposed the proposal during the four-month comment period allowed by the United States Department of Agriculture (USDA). The unexpected number of people opposing the proposal was enough to make the USDA reconsider its position. We last heard that the USDA is expected to redraft and resubmit the proposal for public comment and has promised that genetic engineering will not be included in the discussion. Hopefully, the proposed definition will exclude irradiation and sewage sludge as well.

We were appalled to learn that in thirteen US states anti-food slander laws (also referred to as "veggie libel laws") have been enacted. These dubious laws enable companies to sue anyone who makes a disparaging statement against a perishable food product which they consider false or which cannot be substantiated by reasonable, reliable scientific inquiry, facts or data. Critics fear the libel laws go too far in protecting agribusiness and will threaten free speech. At a time when people are increasingly concerned about food safety issues, these laws could limit public discussion. People who are not part of the scientific community may be hesitant to speak out or question practices they consider unsafe. It would have a muzzling effect on consumer groups and critics.

The much publicized Oprah Winfrey case is a good example. On April 16, 1996, the influential and popular talk-show hostess interviewed William Hueston of the US Department of Agriculture, Gary Weber of the National Cattlemen's Beef Association and Howard Lyman, an official of the US Humane Society's Eating With a Conscience Campaign. Oprah devoted the entire episode of her show to mad cow disease and the threat that it posed to cows and people on this side of the Atlantic.

Mr. Lyman, a former Montana cattle rancher, argued that the US Beef Industry's practice of feeding ground-up remains of dead cattle to other cows could lead to a similar tragic epidemic in the US and compared the disease and its incubation period to AIDS, stating that it was already rampant in the United States. He claimed that "100,000 cows per year in the US are fine at night and dead in the morning," and ". . . they are rounded up, ground up and fed back to other cows. If only one of them has mad cow disease, it has the potential to affect thousands." William Hueston and Gary Weber disagreed and said that this practice was perfectly safe. Oprah, however, sided with Lyman stating to her audience, "It has just stopped me cold from eating another hamburger. I'm stopped."

Following the broadcast, cattle prices fell to near ten year lows and a group of irate cattle ranchers filed suit against Oprah Winfrey in June 1996 claiming more than twelve million dollars in losses. What followed was a lengthy, tedious trial and though Oprah won the case, it cost her close to a million dollars. As a result of this experience, Oprah intends to be less outspoken on such issues.

One outcome was that the US Department of Agriculture considered drafting a regulation in January, 1998 which, if adopted, would prohibit the practice of fortifying animal feed with cow renderings. However, this regulation was never passed.

It is time consumers insisted that genetically altered foods be clearly labelled as such and demand that only truly organic foods be permitted to be labelled organic. If governments devoted as much time and money in supporting organic farming as they do in supporting hi-tech multinationals, consumers would have little reason to be concerned. We strongly recommend that consumers urge their government to support and favour organic farming methods for the protection of our environment and our health.

The Cost of "Progress"

Biotech industries have deployed a most powerful and dangerous technology and it is being quickly implemented worldwide. Products are being permitted on the market without having been adequately tested for long-term and unexpected side-effects.

What is more, governments are making hasty decisions which propel the food industry steadily towards self-regulation. As with all businesses, the ultimate goal of multinationals is growth and as such their priorities in producing foods have little in common with those of the average consumer. Their primary interest is in techniques which will reduce the cost of production, enable expanded distribution, reduce spoilage and increase yields and sales. Nutrition and safety considerations will only become priorities when it becomes mandatory to supply relevant test data with the product.

The Union of Concerned Scientists (a body of more than 1,600 respected scientists from around the world, including more than 100 Nobel Laureates in the sciences) believes that genetically engineered foodstuffs can pose significant risks to human health and the environment. They advocate sustainable agriculture as a more intelligent, viable solution.

Physicians and Scientists Against Genetically Engineered Food are demanding "a global moratorium on the release of genetically engineered food products into the environment and on the use of them as food." Concerned experts are seeking an independent public inquiry to investigate the risks and hazards involved. This inquiry would bring together the most current and comprehensive scientific knowledge coupled with social and moral standards. The ultimate goal would be a code of good practice which would govern the testing, evaluation, documentation and control over the proliferation of genetically modified species. Public opposition to genetic engineering and biotechnology has been gaining momentum throughout Canada, Europe and the United States. We were reassured to read in *The Globe and Mail* that Prince Charles had announced that no genetically altered food would ever pass his lips: "That takes mankind into realms that belong to God, and God alone."

Alternatives and Action

The following chart demonstrates the advantages of organic farming over genetic engineering.

Organic Agriculture	**Genetically Engineered Agriculture**
Starts by building healthy soil without chemical contamination using methods that harmonize with nature.	Attempts to control the environment by heavy use of chemical fertilizers, pesticides and herbicides thereby contaminating the environment.
Uses crop rotation thereby reducing the number of pests that may affect any one crop. Different crops are grown from year to year. This method of agriculture diversifies crops and treats soil, water, crops, climate, animals, pests and wildlife as parts of an interrelated whole.	The same crops are planted year after year providing pests with opportunities to build up their numbers. Views pests as an isolated problem to be solved with a pesticide. If pests develop resistance to the pesticide, then new pesticides are used which results in more contamination.
Uses nature's methods to attack insects, e.g., introduction of other predatory insects or birds to kill off the problem insect.	Attempts to solve pest problems by equipping each crop with pesticide genes.
Uses sustainable agricultural practices which can produce crops decade after decade without loss of yields and without water or soil pollution. This is desirable from a human health standpoint.	Poses serious ecological risks due to complex and unpredictable interactions between new genes. Introduction of novel chemical-resistant crops results in generous use of poisonous pesticides and herbicides causing accelerated rates of contamination.

Simply put, food comes from nature. If biotech scientists are permitted to alter the basic makeup of food, we will see the emergence of strange diseases just as we saw with pesticides and herbicides in the past. Science should be used to better understand nature rather than manipulate it. The ability to live sustainably with nature will only occur when we gain that understanding and respect.

The following chart illustrates the arguments made for and against the use of genetic engineering in food production.

Claims by Opponents of Genetic Engineering	Claims by Supporters of Genetic Engineering
Genetic engineering is controlled by powerful, global interests concerned more with expanding their market than humanity or the environment. It does not contribute to better nutrition.	Longer shelf-life for perishable produce means increased flexibility and lower costs for producers, transporters and sellers.
Deregulation and globalization are undermining the effectiveness of unions and governments and are putting the power of multinational corporations beyond the reach of public accountability.	Genetically engineered foods are expected to become a $50 billion business.
Enough perfectly good food can be grown without genetic engineering. Indeed, in many cases, our farmers are being paid *not* to grow crops, or, are being paid to destroy their crops. Manufacturers of gene-altered foods are motivated by commercial reasons – not by nutritional or ecological reasons.	Allows increased use of forests, lands and fisheries by replacing food plants and animals with cell cultures.
Leads to increased use of agro-chemicals and more pollution of soil, water and food. Organic methods are what nature intended.	Plants can be made resistant to pesticides or to insects, viruses, fungi or agro-chemicals.
Genetic engineering will not solve world hunger caused by a century of environmental abuse, agricultural policies and huge income disparities. It is very unlikely the Third World could afford this technology anyway.	The elimination of world hunger.
Genetic engineering causes serious threats to nature through the release of novel organisms. It is a dangerous global experiment for the sake of short-term commercial gain by large corporations.	This technology offers miracle crops that are resistant to drought and disease.

Important changes to our laws often occur with little public advertisement. Consumers do have some sources of information, though they are expensive and not readily accessible. One of the most important is the *Canada Gazette* which is the only official medium for the publication of all legally binding decisions of the Government of Canada. The federal government gives Canadians advance notice of its intentions by pre-publication in the *Gazette* to elicit comments from the general public and businesses who would be affected by changes in proposed regulations.

The inaccessibility of the day-by-day workings of government leaves most people feeling decisions and laws are made in Ottawa without their input. Since the *Gazette* is the method Ottawa uses to inform Canadians of issues that directly effect them and is written and published with tax dollars, it should be provided free of charge to all taxpayers. This will only happen if enough of us demand it. If you are one of the millions of Canadians constantly griping about the inner workings of government, here is your chance to change this situation. Let Ottawa know what you want. They are not mind-readers. Waiting until election day is futile. In order to set the wheels in motion, your MP needs to hear from you now. The *Canada Gazette* is available in most public libraries strictly for consultation purposes. To subscribe to the *Gazette* write to: Canadian Government Publishing, Public Works and Government Services Canada, Ottawa, Canada, K1A 0S9. For more in-depth information on the *Canada Gazette*, visit the Government of Canada's website.

Genetic engineering has already caused a number of deaths and human illness as well as untold damage to the environment. As you have seen, genetic engineering is a complex issue involving consumers, food stores, manufacturers, researchers, government and above all, the earth. Though the saying "Man is the author of his own destruction" comes to mind, we believe it is never too late to take action. Constructive consumer action begins when we consider the overall food situation and determine what impact it will have on our lives and on those who are dear to us. If each one of us does our little bit, it will ultimately have a snowball effect leading to positive change. With that in mind, we leave you with the following Consumer Action Plan:

- Support local organic farmers. The Department of Agriculture keeps a list of those registered in your area and could provide you with their names, addresses and phone numbers.
- Keep as up-to-date as possible on what's happening in this area. Read the *Canada Gazette*, newspapers and magazines and keep

your eyes on the Net. *alive* magazine publishes an article called "Biotech News" each month which is a survey of some of the latest developments in genetic engineering.

- Contact and support groups such as the Campaign To Ban Genetically Engineered Foods, Natural Law Party, 500 Wilbrod St., Ottawa, ON, K1N 6N2; Tel. (613) 565-8517 ; Fax: (613) 565-6546; e-Mail: natural_law@ottawa.com or visit their website at: http://www.natural–law.ca/genetic/geindex.html.
- Subscribe to informative and up-to-date newsletters such as *The Ram's Horn* published monthly by Brewster and Cathleen Kneen. Subscriptions cost approximately $20 and can be ordered from P.O. Box 3028, Mission, B.C. V2V 4J3.
- Join a local Food Awareness group for sharing information on safe and nutritious foods. If there isn't one already in existence in your area, now is a good time to form one. Contact your local health food store, talk to friends and you will be surprised at how many other people are just as concerned as you and how many knowledgeable people you have around you who would be willing to share their knowledge.
- Write to your MP, supermarkets, consumer groups and the media to demand honest labelling and a moratorium on genetically engineered foods until their long-term effects are known.
- Circulate the tear-out petition in the back of this book or write your own. Remember, the more voices industry and government hear, the more quickly this unacceptable situation in our food supply will be stopped.

3

Irradiated Foods

Irradiation is a controversial development in food technology whereby manufacturers expose foods to high-intensity beams of gamma rays in order to kill or render insects incapable of reproducing, to destroy micro-organisms in food, to delay the ripening of fruits and vegetables by changing the biochemistry of their cells and to inhibit sprouting during storage. In Canada, manufacturers are permitted to irradiate

- potatoes and onions
- flour, wheat, and whole wheat flour
- whole or ground spices and dehydrated seasoning preparations

See Division 26 of the *Food and Drugs Act* to learn exactly how Health Canada legislates the use of irradiation in foods. Foods that have been irradiated are required by law to have this symbol, known as the radura, on their label.

The only exceptions are irradiated food ingredients that make up ten per cent or less of the final product. In that case, pre-packaged food such as tomato sauces, sausages and frozen pizzas could contain irradiated herbs and spices without displaying the radura on the label. See the *Food and Drugs Act*, regulation B.01.035 (6).

In the US, the FDA has yielded to pressure from the food industry to relax legislation pertaining to the labelling of irradiated foods. Manufacturers are now permitted to use the same size of print on the package to declare that the food has been irradiated as they use to list their ingredients (we all know how small that print can be). The food industry is now petitioning the FDA for the right *not* to label irradiated foods as such. This is just one more step in the gradual erosion of protective legislation. As Canadian legislation is often influenced by laws in the US, we could soon see similar labelling laws here.

Prior to 1989, the Canadian government considered irradiation a food additive and testing for toxicity was mandatory. Then Ottawa changed the rules and now irradiation of food is called a *process*. According to regulation B.26.005 of the *Food and Drugs Act*, Health Canada does not require new tests to be submitted by manufacturers unless Ottawa feels the previous research is out-dated or inappropriate. In their book *Additive Alert*, Pollution Probe, an independent environmental research team, interprets this regulation to mean that testing for toxic effects is no longer mandatory although food manufacturers are required to supply data to show the food has not been significantly altered nutritionally, physically or chemically. "Significantly" is the key word here. Canadians need to know what is considered a significant change. Are consumers willing to accept alterations in the taste, texture and safety of foods? The most important consideration, however, is the impact of cellular changes in irradiated foods and the effect they may have on the human body.

All food processes such as cooking can produce free radicals in food. Irradiation, however, appears to create more. The free radicals that are unique to the irradiation process are shattered pieces of molecules created by nuclear bombardment. When they combine with other components of foods, they produce new substances that are referred to as unique radiolytic products (URPS). Unique radiolytic products are difficult to identify and track and subsequently have not been adequately tested for mutagenic, carcinogenic and teratogenic effects. The fact that free radicals in food have already been cited as cancer-causing agents throws suspicion on unique radiolytic products. Guaranteeing safety from URPS is virtually impossible because of the unpredictable effects of free radicals and their chance combinations with other substances in foods.

Physicists at Melbourne University discovered the levels of free radicals increased between three and fifty times depending on the food that was irradiated. The human body is very sensitive to free radicals produced by gamma radiation. In fact, these free radicals are capable of causing cancer and premature aging. The effects of eating irradiated foods over a life-time are impossible to predict. Furthermore chromosome damage may not become apparent until later generations. A variety of studies in which animals were fed irradiated foods showed effects such as damage to bones, liver, spleen, kidneys, testicles and ovaries as well as premature death, fewer offspring, higher numbers of still-births, lower birth weight, retarded growth, chromosome damage, tumours and cataracts. The following studies exemplify these effects:

- In the 1970s, the National Institute of Nutrition in India conducted experiments on fifteen malnourished children in order to study the effects of irradiated foods on the human body. The findings were used to explore the feasibility of using stored irradiated grains during food shortages. Irradiated wheat fed to the children produced changes in their white blood cells. Human cells contain a complete set of forty-six chromosomes which hold all genetic information. When the white blood cells are changed, they become what are called polyploid cells, abnormal white blood cells containing two or three complete sets of chromosomes (ninety two or one hundred and thirty-eight chromosomes). Polyploidy, generally a rare condition in human cells, was found to increase with the consumption of irradiated food. Undernourished people are already vulnerable to disease and feeding them irradiated food would cause their health to deteriorate even further.
- In 1987, the Shanghai Institute of Radiation Medicine and the Shanghai Institute of Nuclear Research conducted a study on seventy healthy male and female subjects which confirmed that abnormalities in the chromosomes also developed in healthy adults who ate stored irradiated foods.
- In 1968, the FDA reported that laboratory animals who were fed irradiated foods showed increases in testicular tumors, pituitary cancer, weight loss, shortened life span and reduced fertility.
- Lab rats in the former Soviet Union who were fed irradiated foods had increased rates of kidney and testicular damage.
- In Canada, studies revealed that laboratory animals who ate irradiated foods developed an extra set of chromosomes.
- In West Germany, food irradiation was banned when tests indicated it increased the chance of mutations, reduced fertility, metabolic disturbances, decreased growth rate, reduced resistance to diseases, changes in organ weight and cancer.
- In the 1960s, the US Army and the Atomic Energy Commission were pushing for the use of irradiation on food destined for front-line troops. The high levels of radiation needed to permanently

preserve food made the plan unfeasible as it destroyed food nutrients and produced new chemical substances which were suspected of being highly toxic and possibly carcinogenic. Test animals were severely damaged when fed these irradiated foods and in 1966, the WHO recommended extreme care in their use.

The huge jolts of irradiation results in the destruction of vitamins and other essential nutrients. Researchers claim that irradiation diminishes or destroys vitamins A, C, E, K, B_1, B_2, B_3, B_6, B_{12}, carotene and thiamine. Ironically, proponents of food irradiation suggest that irradiated foods simply be fortified with vitamins or that people take vitamin supplements.

As a radioactive atom disintegrates, it emits gamma rays which are electromagnetic waves, similar to X-rays and light, but of shorter wavelength. They are consequently more energetic and much more penetrating; some can readily pass through thick slabs of concrete. Gamma rays used in food irradiation are derived from Cobalt-60 or Cesium-137. Cobalt-60 can be extracted from cobalt ore but since this is a very expensive process it is normally produced in nuclear reactors by bombarding cobalt with neutrons. Cesium-137, the by-product from the nuclear industry, ranks in the top three of the most hazardous fissionable elements. Nuclear fission is best described as the splitting of the nucleus of a heavy atom into two main parts and is accompanied by the release of nuclear energy.

One of the ways in which the deposit of radiation energy is measured is with gray (Gy) units which define the amount of energy deposited in a given weight of tissue (or object). Although much of the radiation goes directly through the food, a small amount is absorbed. The effect on the food will be determined by the size of the radiation absorbed dose (RAD).

The authors of *Additive Alert* point out that one chest X-ray is equal to one rad which is defined as a measure of radiation. Therefore:

1 RAD = .01 Gy = 1 chest X-ray

The following table shows the amount of radiation permitted in some basic foods, and the equivalent amount in chest X-rays.

Potatoes & Onions	15,000 RADS	=	0.15 kGy	= 15,000 chest X-rays
Wheat, Flour, Whole Wheat Flour	75,000 RADS	=	.075 kGy	= 75,000 chest X-rays
Whole spice, ground spices & dehydrated seasoning preparations	1,000,000 RADS	=	10.00 kGy	= 1,000,000 chest X-rays

Health Canada has given clearance for the use of the international recommendation of up to 1,000,000 RADs on foods. This means that food manufacturers are not required to provide evidence of safety unless they use more than 1,000,000 RADs on a product. We know that for one chest X-ray (one RAD) anyone in the immediate area must stand behind a heavy lead shield to protect against possible cellular damage. One wonders what 1,000,000 RADs does to the biological makeup of foods. Some fruits and vegetables require about 1,000,000 RADs and pork needs about 300,000 RADs.

The Canadian Irradiation Centre, in Laval, Quebec, irradiates up to two million pounds of spices per year, including onion flakes and onion powder. They are primarily a training and demonstration centre for marketing irradiation equipment to countries that irradiate a large portion of their foods. According to Health Canada's Irradiation Consumer Protection Division, foods other than spices are not yet being irradiated here because the process is deemed unacceptable to most Canadians. Manufacturers are hesitant to pour large sums of money into irradiation facilities that might not give a steady return on their investment.

In the United States, the FDA permits manufacturers to irradiate a wide variety of foods. Recent outbreaks of E.coli bacteria in beef has prompted industry and some members of the public to press for irradiation of beef. Scientists have warned that irradiation will not destroy all of the bacteria and therefore cannot be used to replace proper handling and cleanliness in packing plants. The danger is that the use of irradiation could encourage sloppy handling of foods not just by the packers but also by consumers. Furthermore, irradiation will not prevent food from becoming contaminated at a later stage.

Dr. Noel Sommer and the late Dr. Edward Maxie spent years researching irradiation with the support of Atomic Energy of Canada Ltd (AECL) and said there is considerable evidence indicating that some diseases of post-harvested citrus fruits occur because of the cellular injury that happens when they are irradiated. Irradiation lowers the fruit's normal resistance to disease. Researchers also tell us that irradiated strawberries weep when cut; citrus fruits are more sensitive to the cold and foods suffer changes in colour, odour, flavour and texture. Manufacturers may then have to add synthetic colours, flavours or other additives to correct these deficiencies before marketing. As a nation, are we prepared to accept all the effects of irradiation?

We owe it to our children to ensure this toxic nightmare winds down and the best way to do this is to make sure that nuclear technology is only brought to bear when there is no viable alternative.

Many countries have banned the process of food irradiation. Among these are Switzerland, Great Britain, Germany, Sweden, New Zealand, Austria, Romania, Abu Dhabi, the Dominican Republic, Botswana, Ethiopia, Kenya and Tanzania. Let's add Canada to this list.

4

From Lab to Larder: A Case Study of Infant Formula and Infant Food

Safe and nutritious food is basic to survival, not just something we like, but something we need. Many of us have lost touch with this most important commodity and we rarely consider where the food in supermarkets comes from, let alone how it has been produced. We take for granted the availability of out-of-season or exotic produce without questioning how far it may have travelled and we are apt to comparison shop for prices before we consider where the item was grown, what processes were used in its production or what may have been added in the form of food additives. Accordingly, we are contributing to the closure of small local producers who find it difficult to compete with the huge centralized enterprises that now provide us with most of our foods.

The products we buy in the supermarkets influence what will be displayed on their shelves and ultimately we influence what foods are produced and how the foods will be put together.

Consumers should bear this simple principle in mind and act accordingly. There are a number of critical issues that influence the quality of your food and, by extension, your life. Lack of adequate testing of genetically engineered foods, irradiated foods and food additives should be the foremost concern of every consumer. Another serious matter is that it is not mandatory to test the dangers of these additives and processes when combined in a single food. If they can be harmful individually, their combination could have further harmful effects.

Toxicologists, by the very nature of their profession, test chemicals one at a time on laboratory animals. This is not a very dependable approach to researching the toxic effects of the many additives a person could consume in one meal and it doesn't even address genetically engineered and irradiated foods. In the words of Dr. Ross Hume Hall "There are no accepted techniques to test the multitoxicity the average person is exposed to. A simple approach might be to feed a typical

Canadian diet to laboratory animals, but nutrition scientists would consider that to be unscientific. Therefore, such experiments are not done."

Another point to consider is that lab mice and rats have an approximate life span of eighteen months while humans generally live sixty-five years or more eating a plethora of food additives every day. Long-term testing of foods which humans eat over the course of a lifetime has not been carried out. Questions are now being raised about the foods we've been eating for decades and their effect on the quality of our life and general well-being. Many people just don't feel well day after day. They feel depleted of energy and sick. They live in a state of diminished capacity – not working at their best or feeling their best. Obviously, one doesn't need to have cancer to feel the negative effects of these technologies on the human body.

Since it was our concern for our children's health that motivated us to question the widespread use of food additives, we have chosen infant formula and infant foods as a case study. In this section, we will discuss the impact that processed foods can have on children's health and development.

The first food many babies will receive is infant formula. Formula is inexpensive to make, has an indefinite shelf-life and can be sold at high prices. Dr. David Reuben, author of *Everything You Always Wanted To Know About Nutrition* singled out infant formula for special comment and described it as a cheap, "artificial-imitation milk."

The anatomy, physiology and body responses of an infant differ greatly from those of an adult. Not all of the infant's organs are fully developed and most food additives will have a detrimental effect on them. Infants develop cells in their kidneys, brain and muscles very quickly after birth and cells in these tissues do not regenerate. Infants have not yet developed body defenses which are capable of detoxifying harmful additives. The energy needs for proper growth in infants are enormous and interference with these in the form of toxic chemicals can retard growth and development.

In the past, manufacturers of baby formula spent years and millions of dollars advertising that they had improved on the substance that nature had adequately provided since the beginning of time. It is difficult to believe that a man-made mixture of chemicals could replace the nutrients found in human milk. Ironically, manufacturers of formula have recently begun to print on their labels statements such as "Breast milk is best for infants and is recommended for as long as possible. . . ."

Nature has wisely provided what the baby needs by making human milk nutritious and full of antibodies which provide immunity against diseases and allergies. In the first few days after birth, mother's milk contains a substance called colostrum which protects the baby from diseases like poliomyelitis (polio), dysentery, viral infections, influenza and staphylococcus infections. All the antibodies and immunity that the baby's mother has acquired are delivered to the baby through her milk. Furthermore, research has shown that the baby receives what is called diathelic immunity. Should the baby come in contact with an infection the mother has never encountered, the infecting agent passes through the child's saliva and into the mother's breast where her body begins to produce antibodies against this specific infection. The antibodies are then passed back to the baby in subsequent feedings to fight the infection.

Scientists have found that mother's milk changes constantly to give the baby exactly what it needs for optimal growth and development. It is well-documented that one of the primary causes of allergies in infants is the consumption of cow's milk or infant formula. Researchers say that cow's milk will probably give the baby immunity to diseases a calf would encounter but leaves the child defenseless against human sickness.

Dr. Reuben suggests that when baby is ready to begin taking solids, parents should be advised that most processed baby food contains small amounts of fruit or vegetables mixed with modified starch and other ingredients to make a watery product appear thick. He also claims there is a total destruction of enzymes and other nutrients due to the fact that these foods are overcooked. Manufacturers are permitted to add tomato paste, onion powder, flavours, dried egg yolk, modified starches and other ingredients to hide or correct these deficiencies.

In addition, Canada has no regulation restricting the use of cured meats containing nitrite (a dangerous chemical) in baby foods and so we have no assurance that it is not being used now, will not be used in the future, or is not being imported legally now. The WHO states that nitrite and nitrate should on no account be used in baby foods and that babies under the age of six months are particularly sensitive. These chemicals have been known to kill babies due to a disease known as methemoglobinemia which deactivates hemoglobin and shuts off the brain's oxygen supply (*see* Sodium nitrate in the Alphabetical Guide).

Look for products without these additions as babies simply do not need them. Companies can remove chemical food additives from baby food but will do so only if consumers make their dollars count in the supermarket by purchasing the least toxic baby foods. Look for baby food producers that produce simple, nutritious foods or make your own.

A sensible and economical alternative to commercial baby food is to prepare your own. Home-made baby food is far superior nutritionally to commercial brands. When preparing the evening meal, include extra vegetables for the baby. Purée the cooked foods separately in your blender with a bit of liquid (like the water in which the carrots were cooked), until smooth, then pour into an ice cube tray and freeze. Once frozen, empty the cubes into a freezer bag, label them and keep them in your freezer. The cost is minimal and you can have up to a week's supply with the right amounts for a meal. Please don't add salt, sugar or spices since your baby simply does not need any more than nature has already provided in whole foods. Also, avoid making more than a week's supply at a time, as the nutritional value will decline after this point.

Children are especially vulnerable to the vast array of chemicals in food. Many children today are exhibiting symptoms associated with multiple chemical sensitivity. Within an hour or so of ingesting foods with colours and flavours, some children become restless, agitated and disruptive. Aggressive behaviour and the inability to concentrate can be manifested over time in lowered mental development and behavioral problems (see Artificial colour and Artificial flavour in the Alphabetical Guide). Barbara Griggs in her excellent book, *The Food Factor*, states that in the US "an estimated ten million children are either hyperactive or else locked into the silent world of autism." All too often, these children are misunderstood by parents, teachers and anyone who comes in contact with them.

We are not only seeing increased behavioural problems in children but also increased incidences of childhood arthritis, cancer and other degenerative and debilitating conditions. Childhood cancer is reported to have escalated nearly thirty-two percent in the last thirty-five years and many conscientious doctors and scientists believe this percentage will climb. It is our belief that this is the result of the continual assault on the human body of chemical food additives.

The *Food and Drugs Act* doesn't contain regulations which specifically outline how infant formula can be manufactured. Human milk substitute is another name for infant formula and is covered in Division 25 of the *Act*. Nowhere in this division, however, could we find a regulation specifying that a manufacturer must provide results of animal testing on the new product for carcinogenic, mutagenic and teratogenic effects. There are regulations that govern which ingredients manufacturers may use in producing new human milk substitutes (*see* the table Additives Permitted In or Upon Infant Formula listed below). These regulations say that a manufacturer may mix any combination of these ingredients in their new product. It does not stipulate that manufacturers must have the product tested for a number of years to see how lab animals would survive this assault of chemicals. It would be more reassuring to see this type of testing done rather than use human babies as guinea pigs.

In order to show the range of chemicals which can end up in infant formula, we will discuss regulation B.25.062 which is a general guideline for the manufacture of infant food and formula in Canada. Then we will reproduce a label from a can of infant formula and point out how some of the ingredients could contain more additives and possibly be genetically engineered.

B.25.062.

(1) Subject to subsection (2), **no person shall sell a food that is labelled or advertised for consumption by infants if the food contains a food additive.**

(2) **Subsection (1) does not apply to**

(a) **bakery products that are labelled** or advertised **for consumption by infants**;

(b) ascorbic acid used in cereals containing banana that are labelled or advertised for consumption by infants;

(c) **soybean lecithin used in rice cereal** labelled or advertised for consumption by infants;

(d) citric acid used in foods that are labelled or advertised for consumption by infants;

(e) **infant formula that contains the food additives set out in Tables IV and X** to Section B.16.100 for use in infant formula; or

(f) **infant formula that contains ingredients manufactured with food additives set out in Table V** to section B.16.100.

In the first line of B.25.062, Ottawa tells us that no additive is permitted in infant food. Then we are told that subsection 1 does not apply to bakery products intended for infants. When we peruse Division 13 of the *Food and Drugs Act*, we discover that bakery products can contain a great many additives. In order to fully understand this, you will need to look at an example of a bakery product more closely, for instance Bread in B.13.021. There you will find a long list of ingredients whose components are not always declared, such as egg yolk. Now, turn to B.22.035 to see what actually goes into the production of egg yolks. If you follow this procedure with each ingredient that is used to make bread, you will see how one bakery product could end up containing hundreds of food additives. Furthermore, if you consider that several crops such as wheat, corn and soyabean are now being genetically engineered, you can see how this technology could easily be included in these products. Now let's consult the Irradiation table in Division 26 which shows that wheat and wheat flour are permitted to be irradiated, so at some point in time, these could also enter infant food.

At this point, if we return B.25.062 (above) and re-read subsection (c) we should question whether the soyabean lecithin is taken from a genetically engineered soyabeans.

If we examine (e) in this same regulation, we are told that infant formula may contain some of the food additives which are listed in Table IV (Emulsifying, Gelling, Stabilizing and/or Thickening Agents) and Table X (pH-Adjusting Agents, Acid-Reacting Materials and/or Water Correcting Agents). For clarity, we have gone through these tables and extracted the additives that are permitted in infant formula.

We have listed these tables below to give you an idea of the numbers involved.

Additives Permitted In or Upon Infant Formula

Acetylated tartaric acid esters of mono and diglycerides	Lecithin
	Monoglycerides
Algin	Mono and diglycerides
Alginic acid	Potassium alginate
Ammonium alginate	Potassium bicarbonate
Ammonium carrageenan	Potassium citrate
Calcium alginate	Potassium carrageenan
Calcium carrageenan	Potassium hydroxide
Calcium citrate	Phosphoric acid
Calcium hydroxide	Sodium alginate
Carrageenan	Sodium bicarbonate
Citric acid	Sodium carrageenan
Guar gum	Sodium citrate
Hydrochloric acid	Sodium hexametaphosphate
Irish Moss gelose	Sodium hydroxide

Finally, if you re-read (f) in regulation B.25.062, you will notice that infant formula is also permitted to contain even more additives which can be found in Table V (enzymes, page 288, in the *Food and Drugs Act*). You will see that some of the enzymes may be taken from genetically engineered sources.

Once you have examined this regulation in detail, you will begin to understand that there is good reason to question the use of hidden additives and the way the *Food and Drugs Act* is written.

Now let's take a close look at the label of a jar of infant formula that is typical of those being purchased by parents every day.

Infant Formula (Concentrate)

Ingredients: Water, corn syrup, sucrose, skim milk powder, coconut oil, soy oil, modified corn starch, calcium phosphate tribasic, potassium citrate, potassium chloride, mono-and diglycerides, soy lecithin, magnesium chloride, maltodextrin, L-methionine, sodium chloride, choline bitartrate, ascorbic acid, carrageenan, taurine, sodium citrate, ferrous sulphate, dl-a-tocopheryl acetate, zinc sulphate.

If we examine regulation B.18.018 (below), we find that syrup may be made up of several other ingredients. By looking up each of these in the Alphabetical Guide, we soon see that corn syrup is not as innocuous as we may have thought.

B.18.018. (S). (Naming the source of the glucose) **Syrup**
(a) shall be glucose;
(b) **may contain**
  (i) **a sweetening agent**,
  (ii) **a flavouring preparation**,
  (iii) sorbic acid,
  (iv) sulphurous acid or its salts,
  (v) salt, and

Several oils are included in the list of ingredients for infant formula. In B.09.001 (below), we see that vegetable fats and oils may contain emulsifying agents, Class IV preservatives and an antifoaming agent. Furthermore, vegetables such as corn and soyabeans, from which these fats and oils have been taken, could be genetically engineered.

> B.09.001. (S). Vegetable fats and oils shall be fats and oils obtained entirely from the botanical source after which they are named, shall be dry and sweet in flavour and odour and with the exception of olive oil, may contain emulsifying agents, Class IV preservatives, an antifoaming agent and B-carotene in a quantity sufficient to replace that lost during processing, if such an addition is declared on the label.

Modified corn starch also appears on this label. We have reproduced Table XIII (below), a list of chemicals which can be used in or upon starch and suggest you read about modified starch in the Alphabetical Guide to see why we feel this item has no place in infant formula. The remaining additives listed on the label of infant formula can also be looked up in the Alphabetical Guide. It is important to remember that it is not only their individual use that is of concern but the unknown result of combining so many of these additives in a single product.

Additives Permitted In or Upon Starch	
Acetic Anhydride	Propylene Oxide
Adipic Acid	Sodium Acetate
Aluminum Sulphate	Sodium Bicarbonate
Epichlorohydrin	Sodium Carbonate
Hydrochloric Acid	Sodium Chlorite
Hydrogen Peroxide	Sodium Hydroxide
Magnesium Sulphate	Sodium Hypochlorite
Nitric Acid	Sodium Trimetaphosphate
Octenyl Succinic Anhydride	Sodium Tripolyphosphate
Peracetic Acid	Succinic Anhydride
Phosphorus Oxychloride	Sulphuric Acid
Potassium Permanganate	

Before we finish, let's consider a number of other ingredients commonly found on baby food labels. For the most part, they are ingredients that sound basic and pure but which in reality add more additives to the product. For example, does the cream in B.08.075 look like the cream our grandmothers used?

Cream	B.08.075. (S)
Evaporated Skim Milk or Concentrated Skim Milk	B.08.011. (S)
Skim Milk Powder	B.08.014. (S)
Liquid Yolk, Dried Yolk or Frozen Yolk	B.22.035. (S)
Shortening, other than butter or lard	B.09.011. (S)
Monoglycerides and Diglycerides	B.09.012. (S)
Canned Vegetable	B.11.002. (S)
Tomatoes or Canned Tomatoes	B.11.005. (S)
Tomato Paste	B.11.009. (S)
Tomato Purée	B.11.012. (S)
Canned Fruit (naming the fruit)	B.11.101. (S)
Fruit Juice (naming the fruit)	B.11.120. (S)
Apple Juice	B.11.123. (S)
Orange Juice	B.11.128. (S)
Concentrated Juice (naming the fruit)	B.11.130. (S)
Rice	B.13.010. (S)
Stew (naming the meat)	B.14.065. (S)

The next time you are shopping, jot down a few of the ingredients listed on the labels. Then come back to this book and follow the steps we have just led you through. We think you will be shocked to find that some apparently pure products are actually saturated with chemicals. It is unacceptable to allow so many additives to be combined in a single product. It shouldn't be difficult for Health Canada to ban the known toxins. Rather than undertake new testing on additives whose toxicity has already been demonstrated, why not start by banning them from the food supply. Then we might have a little confidence in our legislators and our foods.

Cleaning up the food supply in this way is an onerous enough task without further complicating it with new technologies. Researchers the world over are concerned that genetic engineering and irradiation will add unknown and uncontrollable toxic by-products to our foods. It is impossible to predict or prevent the toxins which will result from combining them with food additives.

Here is an interesting thought: what if we are wrong and genetic engineering, irradiation and chemical additives aren't dangerous? Our error would still help, not injure, as these technologies don't contribute to good nutrition anyway.

But what if we are right?

5

Excerpts From the Canadian *Food and Drugs Act*

The following excerpts from the *Food and Drugs Act* demonstrate the wide range of additives permitted in food. Please note, however, that not all the additives contained in a product are necessarily listed on the label. For example, Regulation B.01.009 lists thirty-six ingredients whose components need not be listed on the label. Most of these ingredients contain a wide variety of chemicals in their make-up. Hence, labels do not necessarily inform you of all the chemicals in your foods. It is easy to see how the average person can consume up to 1,800 food additives per day. You will note in these excerpts the use of terms such as anticaking agents, stabilizing agents, Class II preservatives, and so on. The specific additives covered under these headings are listed in Tables of Additives According to Function.

The addition of so many diverse substances to the food supply poses a considerable risk that many of these additives, or their chance combination, will be toxic. These excerpts provide a behind-the-scenes look at the laws that regulate the food supply. Should you wish to review the complete *Act*, it is available for approximately $90.00 from Public Works and Government Services, Publishing, Ottawa, Ontario K1A OS9, Catalogue No.: H41-1-1997E.

We have replicated the *Act* in its original format but on a miniaturized scale. Due to lack of space, we chose only the most common foods and ingredients. Sections of the regulations that were not particularly relevant were omitted. Some sentences have been abbreviated where chemical, Latin or French terms were used, to make the information more easily read. In addition, any reference to the use of these chemicals in drugs and cosmetics has been removed as these were beyond the scope of this book. Thus, we have taken some liberties with the *Act* but have been faithful to Health Canada in the reproduction of Canadian food laws. The majority of the regulations are complete as found in the original work. What is omitted, in no way changes the intent of the *Act*.

The terminology used in these regulations will often give you the sensation of walking through a maze and you may be unsure of their intent. We read some regulations many times and found that if we bolded certain words we were better able to understand what they were saying. As this worked for us, we have included the bold print here to help you negotiate this maze with better comprehension. The *Act* is actually very simple to use even though it looks daunting. All you have to do is consult the following Table of Contents, choose the food or subject you would like to know more about and be prepared for some real eye-openers.

Ultimately, these sections of the *Food and Drugs Act* should enable you to recognize the inadequacies of legislation that is intended more to facilitate food manufacturers than to ensure the safety of the food supply. We consider this unconscionable and unjustifiable. You be the judge. Should you agree, we urge you to contact your MP and demand changes such as better access to the *Act*, more accountability of decision makers and a review of the *Act* by independent parties. YOU have the power to make these changes happen. Good reading!

Table of Contents of the *Food and Drugs Act*

Interpretations, Regulations and Administration

1. This Act may be cited as the *Food and Drugs Act*.

INTERPRETATION

2. In this Act,
"advertisement" includes any representation by any means whatever for the purpose of promoting directly or indirectly the sale or disposal of any food;

"Department" means the Department of Health;

"food" includes any article manufactured, sold or represented for use as food or drink for human beings, chewing gum, and **any ingredient that may be mixed with food for any purpose whatever;**

"inspector" means any person designated as an inspector for the purpose of the enforcement of this Act;

"label" includes any legend, work or mark attached to, included in, belonging to or accompanying any food;

"Minister" means the Minister of Health;

"package" includes anything in which any food is wholly or partly contained, placed or packed;

"prescribed" means prescribed by the regulations;

"sell" includes offer for sale, expose for sale, have in possession for sale and distribute, whether or not the distribution is made for consideration;

GENERAL

Authors' Note: *The laws and penalties concerning the adulteration of food were changed in May 1997 for the first time in over twenty years. The vast majority of our food supply is adulterated, contains harmful substances and is unfit for human consumption. The penalties for infractions of these laws are so minimal that multinational food companies are apt to simply pay them and carry on business as usual. If more realistic penalties were enacted and then applied in a consistent manner, the result would be an unadulterated food supply.*

FOOD

4. No person shall sell an article of food that

(a) **has in or upon it any poisonous or harmful substance;**
(b) is unfit for human consumption;
(c) consists in whole or in part of any filthy, putrid, disgusting, rotten, decomposed or diseased animal or vegetable substance;
(d) **is adulterated**; or
(e) was manufactured, prepared, preserved, packaged or stored under unsanitary conditions.

5. (1) **No person shall** label, package, treat, process, **sell or advertise any food in a manner that is false, misleading or deceptive** or is likely to create an erroneous impression **regarding its character**, value, quantity, **composition**, merit **or safety**.

REGULATIONS

30.

(1) **The Governor in Council may make regulations** for carrying the purposes and provisions of this *Act* into effect, and, in particular, but without restricting the generality of the foregoing, may make regulations

(a) **declaring that any food** or class of food **is adulterated if any** prescribed **substance** or class of substances is present therein or **has been added thereto** or extracted or omitted therefrom;
(b) **respecting**

(i) the labelling and packaging and the offering, exposing and advertising for sale of food,
(ii) the size, dimensions, fill and other specifications of packages of food,
(iii) the sale or the conditions of sale of any food, and
(iv) **the use of any substance as an ingredient in any food, to prevent the** purchaser or **consumer** thereof **from being deceived or misled in respect of the** design, construction, performance, intended use, quantity, character, value, **composition**, merit **or safety thereof,** or to prevent injury to the health of the purchaser or consumer;

(c) prescribing standards of composition, strength, potency, purity, quality or other property of any article of food,
(e) respecting the method of manufacture, preparation, preserving, packing, storing and testing of any food, in the interest of, or for the prevention of injury to, the health of the purchaser or consumer;
(f) requiring persons who sell food, to maintain such books and records as the Governor in Council considers necessary for the proper enforcement and administration of this Act and the regulations;
(j) **exempting any food, from all or any of the provisions of this Act and prescribing the conditions of the exemption;**
(k) prescribing forms for the purposes of this Act and the regulations;
(l) providing for the analysis of food, other than the purposes of this Act and prescribing a tariff of fees to be paid for that analysis;
(m) adding anything to any of the schedules, in the interest of, or for the prevention of injury to, the health of the purchaser or consumer, or deleting anything therefrom;

OFFENCES AND PUNISHMENT

31.1
Every person who contravenes any provision of this Act or the regulations, as it relates to food, **is guilty of an offence and liable**
(a) **on summary conviction, to a fine not exceeding $50,000**

or to imprisonment for a term not exceeding six months or to both; **or**

(b) **on conviction** by indictment, **to a fine not exceeding $250,000 or to imprisonment for a term not exceeding three years or to both.**

ADMINISTRATION

General

A.01.001.
These regulations may be cited as the Food and Drugs Regulations.

A.01.002.
These regulations, where applicable, prescribe the standards of composition, strength, potency, purity, quality or other property of the article of food or drug to which they refer.

Interpretation

A.01.010.
In these regulations
"acceptable method" means a method of analysis or examination designated by the Director as acceptable for use in the administration of the Act and these Regulations;

"Act" means the *Food and Drugs Act*;

"Director" means the Assistant Deputy Minister, Health Protection Branch, of the Department;

"inner label" means the label on or affixed to an immediate container of a food;

"manufacturer" or "distributor" means a person, including an association or partnership, who under their own name, or under a trade, design or word mark, trade name or other name, word or mark controlled by them, sells a food;

"official method" means a method of analysis or examination

designated as such by the Director for use in the administration of the Act and these Regulations;

A.01.011.
The Director shall, upon request, furnish copies of official methods.

Authors' Note: *If you are curious about some of the official methods of analysis used in the Regulations below, write the Director of Health Canada, Tunney's Pasture, Ottawa, Ontario K1A OL2 and ask for a copy of the official method for whichever food regulation you want to know more about. Quote this Regulation Number as your authority for receiving it. After all, A.01.011 is the law.*

DIVISION 1

General

B.01.001.
In this Part
"common name" means, with reference to a food,
- (a) the name of the food printed in boldface type in these Regulations,
- (b) the name prescribed by any other regulation, or
- (c) if the name of the food is not so printed or prescribed, the name by which the food is generally known;

"component" means an individual unit of food that is **combined** as an individual unit of food **with one or more other individual units of food to form an ingredient;**

"flavouring preparation" includes any food for which a standard is provided in Division 10;

"food additive" means any substance the use of **which results,** or may reasonably be expected to result, **in it or its by-products becoming a part of** or affecting **the characteristics of a food, but does not include**
- (a) any nutritive material that is used, recognized, or commonly sold as an article or ingredient of food,

(b) vitamins, mineral nutrients and amino acids, other than those listed in the tables to Division 16,
(c) **spices, seasonings, flavouring preparations, essential oils, oleoresins and natural extractives,**
(d) **agricultural chemicals,** other than those listed in the tables to Division 16,
(e) **food packaging materials** and components thereof, and
(f) **drugs** recommended **for administration to animals that may be consumed as food;**

"ingredient" means an individual unit of food that is **combined** as an individual unit of food **with one or more individual units of food to form an integral unit of food** that is sold as a prepackaged product;

"simulated meat product" means any food that does not contain any meat product, poultry product or fish product **but** that **has the appearance of a meat product;**

"sugars" means all monosaccharides and disaccharides;

"sweetener" means any food additive listed as a sweetener in Table IX to section B.16.100;

"sweetening agent" includes any food for which a standard is provided in Division 18, but does not include those food additives listed in the tables to Division 16;

"unstandardized food" means any food for which a standard is not prescribed in this Part;

"yolk-replaced egg" means a food that
(a) **does not contain egg yolk but contains** fluid, dried or frozen egg albumen or **mixtures thereof,**
(b) is intended as a substitute for whole egg, **and**
(c) **meets the requirements of section B.22.032.**

B.01.002.
Each provision in this Part **in which the symbol [S] appears between the provision number and the name of the food** described

in that provision **prescribes the standard of composition,** strength, **potency, purity,** quality **or** other **property of that food and a provision in which the symbol does not appear does not prescribe a standard for a food.**

Authors' Note: *The following excerpt gives one the impression that all ingredients in a product are listed on the label–subject to regulation B.01.009. "Subject to" are the key words here. If you examine B.01.009 you'll find a long list of common ingredients whose components need not be declared on the label. This is why it is possible for thousands of chemical additives to be used in foods while approximately four hundred (listed in the Tables of Additives According to Function) must be declared on labels. The rationale used by industry and government for excluding these components from food labels is that the ingredients list would be too long and complicated if manufacturers were required to list everything. Perhaps if we demanded realistic labelling, manufacturers would be more inclined to remove many unnecessary and dangerous additives in order to have a more appealing label. Furthermore, if manufacturers put more effort into providing accurate labelling in lieu of eye-catching labels, there would be plenty of room for listing additives. Now that would be progress and positive change!*

B.01.008.

(1) The following information shall be shown grouped together on any part of the label:
   (a) any information required by these Regulations, other than the information required to appear on the principal display panel and the information required by section B.01.007 and B.01.310; and
   (b) where a prepackaged product consists of more than one ingredient, a list of all ingredients, including, subject to section B.01.009, components, if any.

(2) Paragraph (1)(b) does not apply to
   (a) prepackaged products packaged from bulk on retail premises, except prepackaged products that are a mixture of nuts;
   (b) prepackaged individual portions of food that are served by a restaurant or other commercial enterprise with meals or snacks;
   (c) prepackaged individual servings of food that are prepared by a commissary and sold by automatic vending machines or mobile canteens;

(d) prepackaged meat and meat by-products that are barbecued, roasted or broiled on the retail premises;
(e) prepackaged poultry, poultry meat or poultry meat by-products that are barbecued, roasted or broiled on the retail premises;
(f) bourbon whisky and prepackaged products subject to compositional standards in Division 2; or
(g) prepackaged products subject to compositional standards in Division 19.

(3) **Ingredients shall be shown in descending order of their proportion** of the prepackaged product or as a percentage of the prepackaged product and the order or percentage shall be the order or percentage of the ingredients before they are combined to form the prepackaged product.

(4) **Notwithstanding subsection (3), the following ingredients may be shown at the end of the list of ingredients in any order;**
(a) spices, seasonings and herbs, except salt;
(b) **natural and artificial flavours;**
(c) **flavour enhancers;**
(d) **food additives,** except ingredients of food additive preparations or mixtures of substances for use as a food additive;
(e) vitamins;
(f) salts or derivatives of vitamins;
(g) mineral nutrients; and
(h) salts of mineral nutrients.

(5) Components shall be shown
(a) immediately after the ingredient of which they are components in such a manner as to indicate that they are components of that ingredient; and
(b) in descending order of their proportion of the ingredient.

(6) Notwithstanding paragraph (1)(b) and subsection (5), but subject to section B.01.009, where one or more components of an ingredient are required by these Regulations to be shown in the list of ingredients on the label of a prepackaged product,

the ingredient that contains the components is not required to be shown in the list if all components of that ingredient are listed by their common names with the other ingredients of the product.

(a) in descending order of their proportion of the product, or
(b) as a percentage of the product, the order or percentage, as the case may be, being based
(c) in the case of components, on the total amount of each of the components before they are combined to form ingredients in the product, and
(d) in the case of ingredients, on the amount of each of the ingredients before they are combined to form the product.

(7) Notwithstanding paragraph (1)(b), wax coating compounds and their components are not required to be shown on the label of a prepackaged fresh fruit or fresh vegetable as an ingredient or component thereof.

(8) Notwithstanding paragraph (1)(b), sausage casings are not required to be shown on the label of prepackaged sausages as an ingredient or component thereof.

(9) Notwithstanding paragraph (1)(b), hydrogen, when used for hydrogenation purposes, is not required to be shown on the label of any prepackaged product as an ingredient or component thereof.

(10) Notwithstanding paragraph (1)(b), components of ingredients of a sandwich made with bread are not required to be shown in the list of ingredients on the label of the sandwich.

B.01.009.

(1) **Components of ingredients** or of classes of ingredients **set out in the following table are not required to be shown on a label:**

Ingredients

1. butter
2. margarine
3. shortening
4. lard
5. leaf lard
6. monoglycerides
7. diglycerides
8. rice
9. starches or modified starches
10. breads subject to compositional standards in sections B.13.021 to B.13.029
11. flour
12. soy flour
13. graham flour
14. whole wheat flour
15. baking powder
16. milk subject to compositional standards in sections B.08.003 to B.08.027
17. chewing gum base
18. sweetening agents subject to compositional standards in sections B.18.001 to B.18.018
19. cocoa
20. salt
21. vinegars subject to compositional standards in sections B.19.003 to B.19.007
22. bourbon whisky and alcoholic beverages subject to compositional standards in sections B.02.001 to B.02.134
23. cheese for which a standard is prescribed in Division 8, if the total amount of cheese in a prepackaged product is less than 10 per cent of that packaged product
24. jams, marmalades and jellies subject to compositional standards in sections B.11.201 to B.11.241 when the total amount of those ingredients is less than 5 per cent of a prepackaged product
25. olives, pickles, relish and horse-radish when the total amount of those ingredients is less than 10 per cent of a prepackaged product
26. one or more vegetable or animal fats or oils for which a standard is prescribed in Division 9, and hydrogenated, modified or interesterified vegetable or animal fats or oils, if the total of those fats and oils as are contained in a prepackaged product is less than 15 per cent of that prepackaged product
27. prepared or preserved meat, fish, poultry meat, meat by-product or poultry by-product when the total amount of those ingredients is less than 10 per cent of a prepackaged product that consists of an unstandardized food
28. alimentary paste that does not contain egg in any form or any flour other than wheat flour
29. bacterial culture
30. hydrolyzed plant protein
31. carbonated water
32. whey, whey powder, concentrated whey, whey butter and whey butter oil
33. mould culture
34. chlorinated water and fluorinated water
35. gelatin
36. toasted wheat crumbs used in or as a binder, filler or breading in or on a food product

(2) Subject to subsection (3), **where a preparation or mixture set out in the table to this subsection is added to a food, the ingredients** and components **of the preparation** or mixture **are not required to be shown on the label of that food.**

Preparation/Mixture

1. **food colour preparations**	7. mineral preparations
2. **flavouring preparations**	8. **food additive preparations**
3. **artificial flavouring preparations**	9. **rennet preparations**
4. **spice mixtures**	10. **food flavour-enhancer preparations**
5. **seasoning or herb mixtures**	11. compressed, dry, active or instant yeast preparations
6. vitamin preparations	

(3) **Where a preparation or mixture** set out in the table to subsection (2) **is added to a food, and the preparation** or mixture **contains one or more of the following ingredients** or components, **those ingredients** or components **shall be shown by their common names in the list of the ingredients** of the food to which they are added as if they were ingredients **of that food:**
(a) salt;
(b) **glutamic acid or its salts;**
(c) **hydrolyzed plant protein;**
(d) **aspartame;**
(e) potassium chloride; and
(f) any ingredient or component that performs a function in, or has any effect on, that food.

(4) Notwithstanding subsections (1) and (2) where any of the following components is contained in an ingredient set out in the tables to those subsections, that component shall be shown in the list of ingredients:
(a) peanut oil;
(b) **hydrogenated or partially hydrogenated peanut oil;** and
(c) modified peanut oil.

B.01.010.
(1) In this section, "common name" includes a name set out in column II of the tables to subsection (3).

(2) An ingredient or component shall be shown in the list of ingredients by its common name.

(3) For the purposes of subsection (2),
(a) **the ingredient** or component set out **in column I** of an item of the following table **shall be shown in the list of ingredients by the common name set out in column II** of that item:

Ingredient or Component	Common Name
1. any oil, fat or tallow **described in** section **B.09.002 of Division 9**, except lard, leaf or suet	the name of the meat from which the oil, fat or tallow is obtained plus oil, fat or tallow
2. shortening or margarine containing fats or oils, except shortening or margarine containing coconut oil, palm oil, palm kernel oil, peanut oil or cocoa butter	shortening or margarine containing modified by vegetable oil or marine oil or be the common name of the vegetable, animal or marine oil or fat used
3. shortening or margarine containing coconut oil, palm oil, palm kernel oil, peanut oil or cocoa butter	shortening or margarine modified by the common name of the vegetable oil or fat used
4. meat	the name of the meat
5. poultry meat	the name of the poultry
6. fish	the name of the fish
7. plant protein product	the name of the plant plus protein product
8. **hydrolyzed plant protein** produced by the enzymatic process	hydrolyzed plus the name of the plant plus protein
9. any protein isolate	the name of the source of the protein plus protein or the common name of the protein isolate
10. any meat by-product described in section B.14.003, other than gelatin	the name of the meat plus by-product or the name of the meat plus the name of the meat by-product
11. any poultry meat by-product described in section B.22.003	the name of the poultry plus by-product or the name of the poultry plus the name of the poultry by-product

12. any oil or fat referred to in **section B.09.002** that has been hydrogenated or partially hydrogenated, including tallow, but not including lard	"hydrogenated" plus the name of the meat from which the oil, fat or tallow is obtained, plus oil, fat or tallow
13. any oil or fat referred to in **section B.09.002 of Division 9**, including tallow, **that has been modified** by the complete or partial removal of a fatty acid	modified plus the name of the meat from which the oil, fat or tallow is obtained, plus oil, fat or tallow
14. one or more vegetable fats or oils, except coconut oil, palm oil, palm kernel oil, peanut oil or cocoa butter, **that have been hydrogenated or partially hydrogenated**	hydrogenated vegetable oil or hydrogenated vegetable fat or hydrogenated plus the specific name of the oil or fat
15. coconut oil, palm oil, palm kernel oil, peanut oil or cocoa butter, **that has been hydrogenated or partially hydrogenated**	hydrogenated plus the specific name of the oil or fat
16. one or more marine fats or oils **that have been hydrogenated or partially hydrogenated**	hydrogenated marine oil or hydrogenated marine fat or hydrogenated plus the specific name of the oil or fat
17. one or more vegetable fats or oils except coconut oil, palm oil, palm kernel oil, peanut oil or cocoa butter **that have been modified** by the complete or partial removal of a fatty acid	modified vegetable oil or modified vegetable fat or modified plus the specific name of the oil or fat
18. coconut oil, palm oil, palm kernel oil, peanut oil or cocoa butter **that has been modified** by the complete or partial removal of a fatty acid	modified plus the specific name of the oil or fat
19. one or more marine fats or oils **that have been modified** by the complete or partial removal of a fatty acid	modified marine oil or modified plus the specific name of the oil or fat

(b) except when one of the ingredients or components set out in column I of the table to this paragraph is shown separately in the list of ingredients by its common name, **all of the ingredients** or components **present in a food set out in column I** of an item of that table **may be shown collectively in the list of ingredients by the common name set out in column II** of that item:

Ingredient or Component	Common Name
1. one or more vegetable fats or oils, except coconut oil, palm oil, palm kernel oil, peanut oil or cocoa butter	vegetable oil or vegetable fat
2. one or more marine fats or oils	marine oil
3. **one or more of the colours listed in Table III of Division 16**, except annatto where used in accordance with paragraph B.14.031(i) or subparagraph B.14.032(d)(xvi)	**colour**
4. One or more substances prepared for their flavouring properties and produced from animal or vegetable raw materials or from food constituents derived solely from animal or vegetable raw materials	flavour
5. **one or more substances prepared for their flavouring properties and derived in whole or in part from components obtained by chemical synthesis**	**artificial flavour, imitation flavour or simulated flavour**
6. one or more spices, seasonings or herbs except salt	spices, seasonings or herbs
7. any of the following in liquid, concentrated, dry, frozen or reconstituted form, namely, butter, buttermilk, butter oil, milk fat, cream, milk, partly skimmed milk, skim milk and any other component of milk the chemical composition of which has not been altered and that exists in the food in the same chemical state in which it is found in milk	milk ingredients

7.1	any of the following in liquid, concentrated, dry, frozen or reconstituted form, namely, calcium-reduced skim milk (obtained by the ion-exchange process), casein, caseinates, cultured milk products, milk serum proteins, ultrafiltered milk, whey, whey butter, whey cream and any other component of milk the chemical state of which has been altered from that in which it is found in milk	modified milk ingredients
7.2	one or more ingredients or components set out in item 7 combined with any one or more ingredients or components set out in item 7.1	modified milk ingredients
8.	**any combination of disodium phosphate, monosodium phosphate, sodium hexametaphosphate, sodium tripolyphosphate, tetrasodium pyrophosphate and sodium acid pyrophosphate**	**sodium phosphate or sodium phosphates**
9.	one or more species of **bacteria**	**bacterial culture**
10.	one or more species of **mould**	**mold culture or mould culture**
11.	preparation containing **rennin**	**rennet**
12.	milk coagulating **enzymes**	**microbial enzyme**
13.	one or more substances the function of which is to impart flavour and that are obtained solely from the plant or animal source after which the flavour is named	the name of the plant or animal source plus the word "flavour"
14.	toasted wheat crumbs made by cooking a dough prepared with flour and water, which may be unleavened or chemically or yeast leavened, and which otherwise complies with the standard prescribed by section B.13.021 or B.13.022	toasted wheat crumbs

15.	that portion of chewing gum, other than the coating, that does not impart sweetness, flavour or colour	gum base
16.	sugar, liquid sugar, invert sugar or liquid invert sugar, singly or in combination	sugar
17.	glucose syrups and isomerized glucose syrups, singly or in combination, where the fructose fraction does not exceed 60 per cent of the sweetener on a dry basis	glucose-fructose
18.	glucose syrups and isomerized, glucose syrups, singly or in combination, where the fructose fraction exceeds 60 per cent of the sweetener on a dry basis	fructose syrup
19.	sugar or glucose-fructose, singly or in combination	sugar/glucose-fructose
20.	water to which carbon dioxide is added	carbonated water
21.	**one or more of the following food additives, namely, potassium bisulphite, potassium metabisulphite, sodium bisulphite, sodium metabisulphite, sodium sulphite, sodium dithionite, sulphurous acid and sulphur dioxide**	**sulphiting agents or sulphites**
22.	demineralized water or water otherwise treated to remove hardness or impurities, or fluoridated or chlorinated water	water
23.	wine vinegar, spirit vinegar, alcohol vinegar, white vinegar, grain vinegar, malt vinegar, cider vinegar or apple vinegar, singly or in combination	vinegar

(4) Notwithstanding subsection (2) and subsections B.01.008(5), **where a food contains ingredients of the same class, those ingredients may be shown by a class name if**

(a) **they consist of more than one component** and are not listed in the table to subsection B.01.009(1); and
(b) their components are shown
    (i) immediately after the class name of the ingredients of which they are components, in such a manner as to indicate that they are components of the ingredients, and
    (ii) in descending order of their collective proportion of those ingredients.

B.01.033.
(1) **Except in the case of infant formula** or a formulated liquid diet, **no person shall sell a food** represented in any manner **as containing hydrolyzed or partially hydrolyzed collagen, hydrolyzed or partially hydrolyzed gelatin or hydrolyzed or partially hydrolyzed casein unless the label carries the following statement** on the principal display panel in the same size type used for the common name:

"CAUTION, DO NOT USE AS SOLE SOURCE OF NUTRITION."

Irradiation Labelling

Authors' Note: *Doesn't the* ***irradiation symbol*** *(reproduced on the following page) give you the impression of wholesomeness? This stylized flower is the symbol industry uses to denote that their product has been irradiated. In our opinion, this symbol is misleading and should be replaced with the word "irradiated" so that consumers would at least be able to choose irradiated food over non-irradiated food if they so desired.*

According to the Irradiation Consumer Protection Division, the only foods being irradiated in Canada at this time are certain spices. Since the regulations are all in place, more foods could be irradiated at any time, so you need to watch for the Radura symbol. As well, the United States is irradiating foods such as strawberries, apples, pork and poultry and we are uncertain as to how much, if any, is being imported into Canada or how well these foods are labelled. See *Division 26 for the regulations pertaining to irradiation and see this Division for labelling regulations.*

B.01.035.
(1) Subject to subsection (8), **where an irradiated food referred to**

in column I of the table to Division 26 is offered for sale as a prepackaged product, **the principal display panel of the label applied to the package shall carry the symbol described in subsection (5).**

(2) Where an irradiated food referred to in column I of the table to Division 26 is not a prepackaged product and is offered for sale, a sign that carries the symbol described in subsection (5) shall be displayed immediately next to the food.

(3) The symbol required pursuant to subsection (1) or (2) shall appear in close proximity on the principal display panel referred to in subsection (1) or on the sign referred to in subsection (2) to one of the following statements or a written statement that has the same meaning:
- (a) "treated with radiation;"
- (b) "treated by irradiation;" or
- (c) "irradiated."

(4) No person shall sell a food referred to in column I of the table to Division 26 that has been irradiated in the manner set out in subsection B.26.003 (2) unless the requirements of subsections (1) to (3) are met.

(5) For the purposes of subsections (1) to (3), the symbol that indicates the irradiated food shall
- (a) have an outer diameter
  - (i) in the case referred to in subsection (1), equal to or greater than the height of the numerical quantity prescribed by section 14 of the Consumer Packaging and Labelling Regulations for the declaration of net quantity of the package, and
  - (ii) in the case referred to in subsection (2), not less than 5 cm; and
- (b) be in the following form

(6) Notwithstanding subsection B.01.009(1), any food referred to in column I of the table to Division 26 that is an ingredient or component of a prepackaged product and that has been irradiated shall, if the food constitutes ten per cent or more of the prepackaged product, be included in the list of ingredients and preceded by the statement **"irradiated."**

(7) The label attached to a shipping container containing any food referred to in column I of the table to Division 26 that has been subjected to the maximum permitted absorbed dose set out in column IV of that table shall carry the statement required by subsection (3) and the statement "Do not irradiate again."

(8) Where a shipping container constitutes the package of the prepackaged product, the label attached to the shipping container shall carry the statement required by subsection (7) but need not carry the symbol required by subsection (5).

(9) **Any advertising of an irradiated food** referred to in column I of the table to Division 26 **shall identify the food as having been irradiated.**

(10) The statements referred to in subsections (3) and (6) to (8) shall be in both official languages in accordance with subsection B.01.012(2).

Authors' Note: *Our understanding of the following Regulation (B.01.042) is that foods fall into two basic categories – standardized and unstandardized. A standardized food is a food for which Health Canada has assigned certain additives to be used in specific amounts and for specific purposes. A standardized food is represented in the regulations by a capital letter (S) after the regulation number.*

If the regulation does not have an (S) after it, this indicates there is no set standard for manufacturers to follow when producing that food. Unstandardized foods can contain any additives listed in B.16.100. Manufacturers can choose any additive they feel will do the job as long as it is listed in one of the fifteen tables and is used in the amount specified for the purpose designated in that table. Due to lack of space, we were unable to reproduce in the Tables the amounts specified by Health Canada for each additive in foods.

*Once again, we wish to point out that manufacturers are not required to list everything on the label that has been added to produce a specific additive or ingredient (B.01.009). Mono- and diglycerides are perfect examples of this regulation. At one time, manufacturers were told how to put mono- and diglycerides together (*see *B.09.012). Although this regulation was repealed in March 1997, we purposefully left it in to show you the components or ingredients that mono- and diglycerides may contain. You will now see it listed only in B.01.009 which tells you the ingredients that comprise mono- or diglycerides are not required to be listed on the label. This is a good example of how one ingredient (which is made up of many chemicals) can contribute hundreds of chemical additives to one meal without you knowing it.*

Read carefully all of the information we have included from the Food and Drugs Act. *Perhaps you will conclude as we did that the regulations were not written with the intention of protecting the consumer. We feel these regulations merely protect food processors and their trade secrets.*

B.01.042.

Where a standard for a food is prescribed in this Part

(a) **the food shall contain only the ingredients included in the standard for the food;**

(b) each ingredient shall be incorporated in the food in a quantity within any limits prescribed for that ingredient; **and**

(c) if the standard includes an ingredient to be used as a food additive for a specified purpose, **that ingredient shall be a food additive set out in one of the Tables to section B.16.100** for use as an additive to that food for that purpose.

B.01.043.

Subject to section B.25.062, **where a standard for a food is not prescribed in this Part,**

(a) **the food shall not contain any food additives except food additives set out in a Table to section B.16.100** for use as additives to that food for the purpose set out in that Table; and

(b) **each such food additive shall be incorporated in the food** in a quantity **within any limits prescribed for that food and food additive in that Table.**

B.01.044.

Where the limit prescribed for a food additive in a Table to section B.16.100 **is stated to be "Good Manufacturing Practice," the**

amount of the food additive added to a food in manufacturing and processing **shall not exceed the amount required to accomplish the purpose for which that additive is permitted to be added** to that food.

B.01.046.

(1) **A food is adulterated if any of the following substances or classes of substances are present therein or have been added thereto:**
- (a) mineral oil, **paraffin wax or petrolatum** or any preparation thereof;
- (b) coumarin, an extract of tonka beans;
- (c) **non-nutritive sweetening agents;**
- (d) cotton seed flour that contains more than four hundred and fifty parts per million of free gossypol;
- (e) fatty acids and their salts containing chick-edema factor or other toxic factors;
- (f) dihydrosafrole;
- (g) isosafrole;
- (h) oil of American sassafras;
- (i) oil of Brazilian sassafras;
- (j) oil of camphor sassafrassy;
- (k) oil of micranthum;
- (l) safrole, or
- (m) oil, extract or root of calamus
- (n) nut and nut products that contain more than fifteen parts per billion of aflatoxin;
- (o) **ethylene thiourea;**
- (p) **chlorinated dibenzo-p-dioxins;** or
- (q) cinnamyl anthranilate.

(2) For the purpose of paragraph (1)(n), the aflatoxin content of a nut or nut product shall be calculated on the basis of the nut meat portion.

B.01.047.

Notwithstanding section B.01.046.
- (a) **a food,** other than sausage casing, **is not adulterated by reason only that it contains** 0.3 per cent or less mineral oil, if good manufacturing practice requires the use of mineral oil;

(b) **chewing gum** is not adulterated **by reason** only **that it contains a paraffin wax base;**
(c) **fresh fruits and vegetables**, except turnips, **are not adulterated by reason** only that **they are coated with not more than 0.3 per cent paraffin wax and petrolatum,** if good manufacturing practices require the use of such coating;
(d) **turnips and cheese are not adulterated by reason** only that **they are coated with paraffin wax** in accordance with good manufacturing practice;
(e) **sausage casing is not adulterated by reason** only **that it contains 5 per cent or less mineral oil** by weight, if good manufacturing practice requires the use of mineral oil;
(f) **fish is not adulterated by reason only that it contains parts 20 per trillion or less of 2,3,7,8-tetrachloro-dibenzoparadioxin;**
(g) bakery products and confectionery are not adulterated by reason only that they contain 0.15 per cent or less petrolatum, if good manufacturing practice requires the use of petrolatum; and
(h) a salt substitute is not adulterated by reason only that it contains 0.6 per cent or less mineral oil, if good manufacturing practice requires the use of mineral oil;
(i) **fruits, vegetables and cereals are not adulterated by reason** only **that they contain 0.05 parts per million or less of ethylene thiourea.**

Authors' Note: *The following is a new regulation which was added to the* Food and Drugs Act *on July 3, 1997. As this regulation is so complex, we have provided our own interpretation following this regulation.*

B.01.056.

(1) Notwithstanding sections B.01.042 and B.01.043, **where a food does not comply with the requirements of these Regulations, the manufacturer** or distributor of the food, or of a food additive, agricultural chemical, veterinary drug, vitamin, mineral nutrient or amino acid present in the food, **may make a written request to the Director**
(a) **for the exemption of the food from** the application, in whole or in part, of **the requirements relating to it under these Regulations;**

(b) **for a Notice of Interim Marketing Authorization concerning the food** that confirms the exemption; **and**
(c) **that these Regulations be amended.**

(2) For the purpose of this section, **a food does not comply with the requirements of these Regulations if**
(a) **any of the following is present in the food:**
(i) **a food additive that is**
(A) **set out in** column I of an item of **any of the tables to section B.16.100 in an amount that exceeds the maximum level of use** set out in column III or IV of that item **in respect of that food, or**
(B) **not included in the standard for that food** in these Regulations,
(ii) **an agricultural chemical**, or any of its derivatives, set out in column I of an item of Table II to Division 15 of Part B, **in an amount that exceeds the maximum residue limit** set out in column III of that item,
(iii) **a veterinary drug** set out in column I of an item of Table III to Division 15 of Part B in an amount **that exceeds the maximum residue limit** set out in column III of that item,
(iv) **an ingredient in a form other than the form described in the standard for that food** in these Regulations, **or**
(v) **a** vitamin, mineral nutrient or **amino acid**
(A) that is not set out in the Table to Section D.03.002, or
(B) **the level of use of which is at variance with the level permitted under these Regulations;**
(b) **the food**
(i) **is not set out in** column IV of **Table II to Division 15** of Part B **and contains an agricultural chemical** or any of its derivatives set out in column I of **that Table,**
(ii) **is not set out in** column IV of **Table III to Division 15** of Part B **and contains a veterinary drug set out in** column I of **that Table, or**
(iii) **is not set out in** column II of **any of Tables I to IV and VI to XV or in** column III of **Table V to section B.16.000, and contains a food additive set out in** column I of **that Table.**

(3) **The request** referred to in subsection (1) **shall be accompanied by the following information:**
   (a) **the common name and description** of the food;
   (b) **the reasons for which the exemption**, the Notice of Interim Marketing Authorization **and an amendment to these Regulations are required;**
   (c) **a description of every proposed variation from the requirements of these Regulations;**
   (d) **adequate data, including results of tests and scientific analysis, that demonstrate that the use of the food will not be detrimental to the health of the purchaser** or user; and
   (e) **where the request relates to the addition of a** vitamin, mineral nutrient or **amino acid to the food, a statement, with supporting documentation, as to the consistency of the request with the General Principles** for the Addition of Essential Nutrients to Foods **adopted by the Joint Food and Agriculture Organization of the United Nations/World Health Organization Codex Alimentarius Commission** and published in the Codex Alimentarius (Rome, 1996), as amended from time to time.

(4) **Where the Director determines after examining the request and the information submitted under subsection (3) that use of the food will not be harmful to the health of the purchaser or user, the Director shall exempt the food from the application**, in whole or in part, of the requirements relating to it under these Regulations **and issue a conditional or unconditional Notice of Interim Marketing Authorization relating to the food. The Director shall also indicate his or her intention to recommend that these Regulations be amended in relation to the food.**

(5) **The Notice of Interim Marketing Authorization issued** under subsection (4) **shall set out**
   (a) **The common name and description** of the food;
   (b) **the reasons for which the exemption is established** and the Notice of Interim Marketing Authorization issued; and
   (c) **such of the following as are applicable to the food:**
      (i) **the maximum residue limit of any agricultural chemical** or any of its derivatives, expressed in parts per million,

(ii) **the maximum level of use of any food additive that the food is permitted to contain,** or have on it, expressed in the applicable units of measurement,
(iii) **the maximum residue limit of any veterinary drug,** expressed in parts per million, and
(iv) **the minimum and maximum limits for the addition of any** vitamin, mineral nutrient or **amino acid**, expressed in the applicable units of measurement.

(6) **The Notice of Interim Marketing Authorization issued under subsection (4) must be published in the *Canada Gazette Part I* and has effect beginning on the date of publication.**

(7) **The Director may, by a notice published in the *Canada Gazette Part I*, revoke an exemption and Notice of Interim Marketing Authorization relating to a food where the Director believes** on reasonable grounds, **after reviewing any information that comes to the attention of the Director, that use of the food is or may be harmful to the health of the purchaser** or user.

(8) **A Notice of Interim Marketing Authorization issued** under subsection (4) **in relation to a food loses its effect on the coming into force of any amendment** to these Regulations **resulting from the Director's recommendation** referred to in that subsection.

Authors' Note: *The following is our interpretation of B.01.056:*

1. *Tells the manufacturer how to apply for Interim Marketing Authorization if their product does not fall under the guidelines of Regulations B.01.042 and B.01.043. This could cover genetically engineered foods, the irradiation of new products and food grown in sewage sludge. It also tells manufacturers they may ask the Director to amend the regulations in the* Act *to include their product.*
2. *Tells manufacturers how to compare their food product to these two regulations (B.01.042 and B.01.043) to see whether or not their product complies with them. Manufacturers will then know if they need to apply for an exemption in order to get their product on the market. It also tells producers if their product contains an agricultural chemical, a veterinary drug or a food additive but the named food does not appear in the list of*

foods in the respective Tables in the Act, *they can apply for an exemption. In addition, it seems they will be able to exceed present chemical residue levels or use any food additive as long as they apply to be exempted from the regulations now in place. Should this be the case, they might as well take the* Food and Drugs Act *and burn it. In our opinion, this regulation seems to allow the Director to give permission to manufacturers to do just about anything they wish and the consumer be damned. This appears to put too much power in the hands of one person, the Director.*

3. *Tells manufacturers what information to send to the Director when they submit their request to have their product exempted from Regulation B.01.042 and B.01.043. They must include testing data demonstrating the safety of the food. This would be fine if it was further stipulated that independent testing would be required to verify the manufacturer's research.*
4. *Tells manufacturers that once the Director has reviewed their documentation and determined that the product will not harm consumers, the Director will accept their request for exemption. It tells manufacturers the Director may even recommend that the regulations be amended so that their food falls under the regulations in the* Food and Drugs Act.
5. *Tells manufacturers how the authorization issued by the Director will be set up. For example, if a manufacturer has applied for permission for a product to contain higher levels of chemical residues than are normally permitted, the Director will tell them what the new maximum level is allowed to be. This means the Director can give permission to have higher levels of agricultural chemical or veterinary drug residues in food than what is presently permitted in the* Act. *As for food additives, this authorization will exempt manufacturers from having to follow the regulations which normally would govern the maximum level used of any specific additive and therefore would set a new level.*
6. *Tells the manufacturer that the Interim Marketing Authorization must be published in the* Canada Gazette, Part I. *It also tells them that they can begin to sell their food product immediately upon publication in the* Gazette.
7. *Tells manufacturers how and when the Director can revoke the exemption he has given their food product and the grounds on which he can do so.*
8. *Tells manufacturers that their Interim Marketing Authorization is no longer needed once the* Food and Drugs Act *has been amended to encompass their food or additive, as a result of the Director's recommendation. It appears to us that the Interim Marketing Authorization is just a method to allow the manufacturer to begin selling his product immediately while Health Canada goes through the tedious motions of having it made into law.*

DIVISION 2

Alcoholic Beverages

B.02.002.

In this Division,
"absolute alcohol" means alcohol of a strength of 100 per cent;

"age" means the period during which an alcoholic beverage is kept under such conditions of storage as may be necessary to develop its characteristic flavour and bouquet;

"alcohol" means ethyl alcohol;

"flavouring" means, in respect of a spirit, any other spirit or wine, domestic or imported, added as a flavouring to that spirit as authorized under the Excise Act;

"grain spirit" means an alcoholic distillate, obtained from a mash of cereal grain or cereal grain products saccharified by the diastase of malt or by other enzymes and fermented by the action of yeast or a mixture of yeast and other micro-organisms, and from which all or nearly all of the naturally occurring substances other than alcohol and water have been removed;

"malt spirit" means an alcoholic distillate, obtained by pot-still distillation from a mash of cereal grain or cereal grain products saccharified by the diastase of malt and fermented by the action of yeast or a mixture of yeast and other micro-organisms;

"molasses spirit" means an alcoholic distillate, obtained from sugar-cane or sugar-cane products fermented by the action of yeast or a mixture of yeast and other micro-organisms, from which all or nearly all of the naturally occurring substances other than alcohol and water have been removed;

"small wood" means wood casks or barrels of not greater than 700 L capacity;

"sweetening agent" means glucose-fructose, fructose syrup or any food for which a standard is provided in Division 18, or any combination thereof.

B.02.010. [S].
Whisky or **Whiskey,** other than Malt Whisky, Scotch Whisky, Irish Whisky, Canadian Whisky, Canadian Rye Whisky, Rye Whisky, Highland Whisky, Bourbon Whisky and Tennessee Whisky.
(a) **shall be** a potable alcoholic distillate, or a mixture of potable alcoholic distillates, **obtained from** a mash of cereal grain or cereal grain products saccharified by the diastase of malt or by other **enzymes** and fermented by the action of yeast or a mixture of yeast **and other micro-organisms; and**
(b) **may contain caramel and flavouring.**

B.02.013. [S].
Malt Whisky
(a) shall be a potable alcoholic distillate, or a mixture of potable alcoholic distillates, obtained by the distillation of a mash of malted grain fermented by the action of yeast or a mixture of yeast and other **micro-organisms**;
(b) shall possess the aroma, taste and character generally attributed to malt whisky; and
(c) **may contain caramel and flavouring.**

B.02.016. [S].
Scotch Whisky shall be whisky distilled in Scotland as Scotch whisky for domestic consumption in accordance with the laws of the United Kingdom.

B.02.017.
No person shall blend or modify in any manner any Scotch whisky that is imported in bulk for the purpose of bottling and sale in Canada as Scotch whisky except by
(a) blending with other Scotch whisky;
(b) the addition of distilled or otherwise purified water to adjust to a required strength; or
(c) **the addition of caramel.**

B.02.020. [S].
(1) **Canadian Whisky, Canadian Rye Whisky or Rye Whisky**

(a) shall
   (i) be a potable alcoholic distillate, or a mixture of potable alcoholic distillates, obtained from a mash of cereal grain or cereal grain products saccharified by the diastase of malt or by other **enzymes** and fermented by the action of yeast or a mixture of yeast and other **micro-organisms,**
   (ii) be mashed, distilled and aged in small wood for not less than three years, in Canada,
   (iii) possess the aroma, taste and character generally attributed to Canadian whisky, and
   (iv) be manufactured in accordance with the requirements of the Excise Act and the regulations made thereunder; and
(b) **may contain caramel and flavouring.**

B.02.022.
(1) Subject to subsection (2), no person shall label, package, sell or advertise any food as **Bourbon Whisky**, or in such a manner that it is likely to be mistaken for Bourbon whisky unless it is whisky manufactured in the United States as Bourbon whisky in accordance with the laws of the United States applicable in respect of Bourbon whisky for consumption in the United States.

(2) A person may modify Bourbon whisky that is imported for the purpose of bottling and sale in Canada as Bourbon whisky by the addition of distilled or otherwise purified water to adjust the Bourbon whisky to a required strength.

B.02.023.
(1) Subject to sections B.02.022 and B.02.022.1, no person shall sell for consumption in Canada any whisky that has not been aged for a period of at least three years in small wood.

(2) Nothing in subsection (1) applies in respect of flavouring contained in whisky, but no person shall sell for consumption in Canada whisky containing any flavouring, other than wine, that has not been aged for a period of at least two years in small wood.

B.02.030 [S].
Rum
(a) shall be a potable alcoholic distillate, or a mixture of potable

alcoholic distillates, obtained from sugarcane or sugarcane products fermented by the action of yeast or a mixture of yeast and other **micro-organisms;** and

(b) may contain
  (i) **caramel,**
  (ii) fruit and other botanical substances, and
  (iii) **flavouring and flavouring preparations.**

B.02.031.

(1) No person shall sell for consumption in Canada any rum that has not been aged for a period of at least one year in small wood.

B.02.033.

No person shall blend or modify in any manner any rum that is imported in bulk for the purpose of bottling and sale in Canada as imported rum except by

(a) blending with other imported rum;
(b) adding distilled or otherwise purified water to adjust the rum to the strength stated on the label applied to the container; or
(c) **the addition of caramel.**

B.02.034.

(1) Notwithstanding section B.02.033, no person shall blend or modify in any manner any rum made from sugar cane products of a Commonwealth Caribbean country that has been distilled and fermented in a Commonwealth Caribbean country and imported in bulk from a Commonwealth Caribbean country for bottling and sale in Canada as Caribbean rum except by
  (a) blending with other rum of a Commonwealth Caribbean country;
  (b) blending with Canadian rum in proportions that result in 1 to 1.5 per cent Canadian rum by volume in the final product,
  (c) adding distilled or otherwise purified water to adjust the rum to the strength stated on the label applied to the container; or
  (d) **adding caramel.**

B.02.041. [S].

Gin, other than Hollands, Hollands Gin, Geneva, Geneva Gin, Genever, Genever Gin or Dutch-type Gin,

(a) shall be a potable alcoholic beverage obtained
  (i) by the redistillation of alcohol from food sources with or

over juniper berries, or by a mixture of the products of more than one such redistillation, or
- (ii) by the blending of alcohol from food sources, redistilled with or over juniper berries, with alcohol from food sources or by a mixture of the products of more than one such blending;

(b) may contain
- (i) other aromatic botanical substances, added during the redistillation process,
- (ii) **a sweetening agent,** and
- (iii) **a flavouring preparation** for the purpose of maintaining a uniform flavour profile; and

(c) may be labelled or advertised as Dry Gin or London Dry Gin if sweetening agents have not been added.

B.02.050. [S].
Brandy, other than Armagnac Brandy or Armagnac, Canadian Brandy, Cognac Brandy or Cognac, Dried Fruit Brandy, Fruit Brandy, Grappa, Lees Brandy and Pomace or Marc,

(a) shall be a potable alcoholic distillate, or a mixture of potable alcoholic distillates, obtained by the distillation of wine; and

(b) may contain
- (i) **caramel,**
- (ii) fruit and other botanical substances, and
- (iii) **flavouring and flavouring preparations.**

B.02.052. [S].
Canadian Brandy

(a) shall be a potable alcoholic distillate, or a mixture of potable alcoholic distillates, obtained by the distillation of wine that has been fermented in Canada; and

(b) may contain
- (i) **caramel,**
- (ii) fruit and other botanical substances, and
- (iii) **flavouring and flavouring preparations.**

B.02.055. [S].
Fruit Brandy

(a) shall be a potable alcoholic distillate, or a mixture of potable alcoholic distillates, obtained by the distillation of

(i) fruit wine or a mixture of fruit wines, or

(ii) a fermented mash of sound ripe fruit other than grapes, or a mixture of sound ripe fruits other than grapes;

(b) may contain

(i) **caramel,**

(ii) fruit and other botanical substances, and

(iii) **flavouring and flavouring preparations;** and

(c) may be described on its label as "(naming the fruit) brandy" if all of the fruit or fruit wine used to make the brandy originates from the named fruit.

B.02.070. [S].
Liqueur or Spirituous Cordial

(a) shall be a product obtained by the mixing or distillation of alcohol from food sources with or over fruits, flowers, leaves or other botanical substances or their juices or with extracts derived by the infusion, percolation or maceration of those botanical substances;

(b) shall have added, during the course of manufacture, a sweetening agent in an amount that is not less than 2.5 per cent of the finished product;

(c) shall contain not less than 23 per cent absolute alcohol by volume; and

(d) may contain

(i) **natural and artificial flavouring preparations, and**

(ii) **colour.**

B.02.080. [S]. Vodka shall be a potable alcoholic beverage obtained by the treatment of grain spirit or potato spirit with charcoal so as to render the product without distinctive character, aroma or taste.

Authors' Note: *Considering the long list of additives permitted in wine and beer, it is strange the government does not require that the ingredients be listed on their labels. Fortunately, organic wines and beer are starting to appear in liquor stores and they are worth seeking out.*

B.02.100. [S].
Wine

(a) shall be the product of the alcoholic fermentation of the juice of the grape;

(b) may have added to it during the course of the manufacture

(i) yeast,
(ii) concentrated grape juice,
(iii) sugar, dextrose, invert sugar, glucose or glucose solids or aqueous solutions thereof,
(iv) yeast foods,
(v) **calcium sulphate** in such quantity that the content of soluble sulphates in the finished wine shall not exceed 0.2 per cent weight by volume calculated as potassium sulphate,
(vi) **calcium carbonate** in such quantity that the content of tartaric acid in the finished wine shall not be less than 0.15 per cent weight by volume,
(vii) **sulphurous acid,** including salts thereof, in such quantity that its content in the finished wine shall not exceed
  (A) 70 parts per million in the free state, or
  (B) 350 parts per million in the combined state, calculated as sulphur dioxide,
(viii) tartaric or citric acid,
(ix) **amylase and pectinase,**
(x) ascorbic or erythorbic acid or salts thereof,
(xi) **anti-foaming agent,**
(xii) any of the following **fining agents: activated carbon, agar-agar, albumen, casein, clay, diatomaceous earth, gelatin, isinglass, potassium ferrocyanide, tannic acid, white of egg and polyvinylpyrrolidone** in an amount that does not exceed 2 parts per million in the finished product,
(xiii) **caramel,**
(xiv) brandy, fruit spirit or alcohol derived from alcoholic fermentation of a food source distilled to not less than 94 per cent alcohol by volume,
(xv) **carbon dioxide,** oxygen or ozone,
(xvi) **sorbic acid** or salts thereof, not exceeding 500 parts per million calculated as sorbic acid, and

(c) prior to final filtration may be treated with
  (i) a strongly acid cation exchange resin in the sodium ion form or
  (ii) a weakly basic anion exchange resin in the hydroxyl ion form

B.02.101.
No person shall sell wine that contains more than 0.13 per cent weight by volume of volatile acidity calculated as acetic acid, as determined by official method FO-2.

B.02.103. [S].
Fruit Wine or **(naming the fruit) Wine** shall be the product of the alcoholic fermentation of the juice of sound ripe fruit other than grape, **and in all other respects shall meet the requirements of the standard for wine as prescribed by section B.02.100.**

B.02.104. [S].
Vermouth shall be wine to which has been added bitters, aromatics or other botanical substances or **a flavouring preparation,** and shall contain not more than 20 per cent absolute alcohol by volume.

B.02.105. [S].
Flavoured Wine, Wine Cocktail, Aperitif Wine shall be wine to which has been added herbs, spices, other botanical substances, fruit juices or **a flavouring preparation,** and shall contain not more than 20 per cent absolute alcohol by volume.

B.02.120. [S].
Cider

(a) shall
- (i) be the product of the alcoholic fermentation of apple juice, and
- (ii) contain not less than 2.5 per cent and not more than 13.0 per cent absolute alcohol by volume; and

(b) may have added to it during the course of manufacture
- (i) yeast,
- (ii) concentrated apple juice,
- (iii) sugar, dextrose, invert sugar, glucose, glucose solids, or aqueous solutions thereof,
- (iv) yeast foods,
- (v) **sulphurous acid,** including salts thereof, in such quantity that its content in the finished cider shall not exceed
  - (A) 70 parts per million in the free state, or
  - (B) 350 parts per million in the combined state, calculated as **sulphur dioxide,**

(vi) tartaric acid and potassium tartrate,
(vii) citric acid,
(viii) lactic acid,
(ix) **pectinase** and **amylase,**
(x) ascorbic or erythorbic acid, or salts thereof,
(xi) any of the following fining agents:
(A) **activated carbon,**
(B) **clay,**
(C) **diatomaceous earth,**
(D) **gelatin,**
(E) **albumen,**
(F) **sodium chloride,**
(G) **silica gel,**
(H) **casein,**
(I) **tannic acid** not exceeding 200 parts per million, or
(J) **polyvinylpyrolidone** not exceeding 2 parts per million in the finished product,
(xii) **caramel,**
(xiii) brandy fruit spirit or alcohol derived from the alcoholic fermentation of a food source distilled to not less than 94 per cent alcohol by volume,
(xiv) **carbon dioxide,**
(xv) oxygen,
(xvi) ozone, or
(xvii) **sorbic acid** or salts thereof, not exceeding 500 parts per million, calculated as sorbic acid.

B.02.130. [S]. Beer
(a) shall be the product of the alcoholic fermentation by yeast of an infusion of barley or wheat malt and hops or hop extract in potable water and shall be brewed in such a manner as to possess the aroma, taste and character commonly attributed to beer; and
(b) may have added to it during the course of manufacture any of the following ingredients:
(i) cereal grain,
(ii) carbohydrate matter,
(iii) salt,
(iv) hop oil,
(v) hop extract, if the hop extract is added to the wort before or during cooking,

(vi) pre-isomerized hop extract,
(vi.1) reduced isomerized hop extract,
(vii) **Irish moss seaweed,**
(viii) **carbon dioxide,**
(ix) **caramel,**
(x) dextrin,
(xi) **food enzymes,**
(xii) **stabilizing agents,**
(xiii) **pH-adjusting** and **water-correcting agents,**
(xiv) **Class I preservatives**
(xv) **Class II preservatives,**
(xvi) **sequestering agent,**
(xvii) **yeast foods,**
(xviii) any of the following **filtering and clarifying agents,** namely, **acacia gum, activated carbon, bentonite, calcium silicate, magnesium silicate, aluminum silicate, cellulose, China clay, Nylon 66, diatomaceous earth, gelatin, isinglass, silica gel, polyvinylpolypyrrolidone** or wood shavings derived from oak, beech, hazelnut or cherry wood.
(xix) **polyvinylpyrrolidone,**
(xx) **ammonium persulphate,**
(xxi) in the case of wort, **dimethylpolysiloxane**, and
(xxii) in the case of mash, **hydrogen peroxide.**

B.02.131. [S]. Ale, Stout, Porter or Malt Liquor

(a) shall be the product of the alcoholic fermentation by yeast of an infusion of barley or wheat malt and hops or hop extract in potable water and shall be brewed in such a manner as to possess the aroma, taste and character commonly attributed to ale, stout, porter, or malt liquor, respectively; and
(b) **may have added to it** during the course of manufacture **any of the ingredients referred to in paragraph B.02.130 (b).**

B.02.132. Where a beer, ale, stout, porter or malt liquor contains the percentage of alcohol by volume set out in column I of an item of the table, the qualified common name or common name set out in column II of that item shall be used in any advertisement of and on the label of the beer, ale, stout, porter or malt liquor.

Percentage Alcohol by Volume		Qualified Common Name or Common Name
1.	1.1 to 2.5	Extra Light Beer, Extra Light Ale, Extra Light Stout, Extra Light Porter
2.	2.6 to 4.0	Light Beer, Light Ale, Light Stout, Light Porter
3.	4.1 to 5.5	Beer, Ale, Stout, Porter
4.	5.6 to 8.5	Strong Beer, Strong Ale, Strong Stout, Strong Porter, Malt Liquor
5.	8.6 or more	Extra Strong Beer, Extra Strong Ale, Extra Strong Stout, Extra Strong Porter, Strong Malt Liquor

B.02.133. [S].

In this Division, "hop extract" means an extract derived from hops by a process employing the solvent

(a) **hexane, methanol, or methylene chloride** in such a manner that the hop extract does not contain more than 2.2 per cent of the solvent used; or

(b) carbon dioxide or ethyl alcohol in an amount consistent with good manufacturing practice.

B.02.134. [S].

(1) In this Division, "pre-isomerized hop extract" means an extract derived from hops by

- (a) the use of one of the following solvents:
  - (i) hexane,
  - (ii) carbon dioxide, or
  - (iii) ethanol; and
- (b) the subsequent isolation of the alpha acids and their conversion to isomerized alpha acids by means of diluted alkali and heat.

(2) For the purposes of paragraph (1)(b), the residues of hexane shall not exceed 1.5 parts per million per per cent iso-alpha acid content of the pre-isomerized hop extract.

B.02.135 [S].
In this Division, "reduced isomerized hop extract" means
(a) tetrahydroisohumulones derived from hops
(i) by isomerization and reduction of humulones (alpha-acids) by means of hydrogen and a catalyst, or
(ii) by reduction of lupulones (beta-acids) by means of hydrogen and a catalyst, followed by oxidation and isomerization; and
(b) hexahydroisohumulones derived from hops by reduction of tetrahydroisohumulones by means of sodium borohydride.

DIVISION 3

Baking Powder

B.03.001.
In this Division, "acid-reacting material" means one or any combination of
(a) lactic acid or its salts;
(b) tartaric acid or its salts;
(c) acid salts of phosphoric acid; and
(d) acid compounds of aluminum.

B.03.002. [S].
Baking Powder shall be a combination of sodium or potassium bicarbonate, **an acid-reacting material, starch** or other neutral material, may contain **an anticaking agent** and shall yield not less than 10 per cent of its weight of carbon dioxide, as determined by official method FO-3.

DIVISION 4

Cocoa and Chocolate Products

Authors' Note: *All of the regulations in this Division were rewritten in May 1997. Unfortunately, the changes have not been for the better as they allow for an increased use of additives in these products. Is it really necessary to so adulterate a much-loved treat of so many children?*

B.04.002. [S].
shall be the seeds of *Theobroma cacao L.* or a closely related species.

B.04.003. [S].
Cocoa Nibs shall be the product prepared by removing the shell from cleaned cocoa beans, of which the residual shell content may not exceed 1.75 per cent by mass.

B.04.005. 1.
Cocoa products may be processed with one or more of the following pH-adjusting or alkalizing agents:

- **(a) hydroxides of ammonia, carbonates of ammonia, bicarbonates of ammonia, hydroxides of sodium, carbonates of sodium, bicarbonates of sodium, hydroxides of potassium, carbonates of potassium or bicarbonates of potassium;**
- **(b) carbonates of magnesium or hydroxides of magnesium; and**
- **(c) carbonates of calcium.**

B.04.006. [S].
Chocolate, Bittersweet Chocolate, Semi-sweet Chocolate or Dark Chocolate

- (a) shall be one or more of the following combined with a sweetening ingredient, namely,
  - (i) cocoa liquor,
  - (ii) cocoa liquor and cocoa butter, and
  - (iii) cocoa butter and cocoa powder;
- (b) Shall contain not less than 35 per cent total cocoa solids, of which
  - (i) not less than 18 per cent is cocoa butter, and
  - (ii) not less than 14 per cent is fat-free cocoa solids; and
- (c) may contain
  - (i) less than 5 per cent total milk solids from milk ingredients,
  - (ii) spices,
  - (iii) flavouring preparations, other than those that imitate the flavour of chocolate or milk, to balance flavour,
  - (iv) salt, and
  - (v) any of the following emulsifying agents, which singly shall not exceed the maximum level of use set out in column III of Table IV to section B.16.100, and in combination shall

not exceed 1.5 per cent by mass of chocolate product, namely,

(A) monoglycerides and mono- and diglycerides,
(B) lecithin and hydroxylated lecithin,
(C) ammonium salts of phosphorylated glycerides,
(D) polyglycerol esters of interesterified castor oil fatty acids, and
(E) sorbitan monostearate.

B.04.008. [S].
Milk Chocolate

(a) shall be one or more of the following combined with a sweetening ingredient, namely,
 (i) cocoa liquor,
 (ii) cocoa liquor and cocoa butter, and
 (iii) cocoa butter and cocoa powder;
(b) Shall contain not less than
 (i) 25 per cent total cocoa solids, of which
  (A) not less than 15 per cent is cocoa butter, and
  (B) not less than 2.5 per cent is fat-free cocoa solids; and
 (ii) 12 per cent total milk solids from milk ingredients, and
 (iii) 3.39 per cent milk fat; and
(c) may contain
 (i) less than 5 per cent total whey or whey products,
 (ii) spices,
 (iii) flavouring preparations, other than those that imitate the flavour of chocolate or milk, to balance flavour,
 (iv) salt, and
 (v) any of the following emulsifying agents, which singly shall not exceed the maximum level of use set out in column III of Table IV to section B.16.100, and in combination shall not exceed 1.5 per cent by mass of chocolate product, namely,
  (A) monoglycerides and mono- and diglycerides,
  (B) lecithin and hydroxylated lecithin,
  (C) ammonium salts of phosphorylated glycerides,
  (D) polyglycerol esters of interesterified castor oil fatty acids, and
  (E) sorbitan monostearate.

B.04.010. [S].
Cocoa or **Cocoa Powder**

(a) shall be the product that
  (i) is obtained by pulverising the remaining material from partially defatted cocoa liquor by mechanical means, and
  (ii) contains not less than 10 per cent cocoa butter; and

(b) may contain
  (i) spices,
  (ii) flavouring preparations, other than those that imitate the flavour of chocolate or milk, to balance flavour,
  (iii) salt, and
  (iv) any of the following emulsifying agents, which singly shall not exceed the maximum level of use set out in column III of Table IV to section B.16.100, and in combination shall not exceed 1.5 per cent by mass of chocolate product, namely,
    (A) monoglycerides and mono- and diglycerides,
    (B) lecithin and hydroxylated lecithin, and
    (C) ammonium salts of phosphorylated glycerides,

DIVISION 5

Coffee

B.05.001. [S].
Green Coffee, Raw Coffee or Unroasted Coffee shall be the seed of *Coffee arabica L.*, *C. liberica Hiern*, or *C. robusta Chev.*, freed from all but a small portion of its spermoderm.

B.05.002. [S].
Roasted Coffee or **Coffee** shall be roasted green coffee, and shall contain not less than 10 per cent fat, and may contain not more than 6 per cent total ash.

B.05.003. [S].
Decaffeinated (indicating the type of coffee).

(a) shall be coffee of the type indicated from which caffeine has been removed and that, as a result of the removal, contains not more than

(i) 0.1 per cent caffeine, in the case of decaffeinated raw coffee and decaffeinated coffee, or
(ii) 0.3 per cent caffeine, in the case of decaffeinated instant coffee; and

(b) may have been decaffeinated by means of **extraction solvents set out in Table XV** to Division 16.

DIVISION 6

Food Colours

B.06.001. In this Division,
"diluent" means any substance other than a synthetic colour present in a colour mixture or preparation;

"dye" means the principal dye and associated subsidiary and isomeric dyes contained in a synthetic colour;

"mixture" means a mixture of two or more synthetic colours or a mixture of **one or more synthetic colours with one or more diluents;**

"official method FO-7" means official method FO-7, Determination of Dye Content of Synthetic Food Colours;

"official method FO-8" means official method FO-8, Determination of Water Insoluble Matter in Synthetic Food Colours;

"official method FO-9" means official method FO-9, Determination of Combined Ether Extracts in Synthetic Food Colours;

"official method FO-10" means official method FO-10, Determination of Subsidiary Dyes in Synthetic Food Colours;

"official method FO-11" means official method FO-11, determination of Intermediates in Synthetic Food Colours;

"official method FO-12" means official method FO-12, Determination of Volatile Matter in Citrus Red No. 2;

"official method FO-13" means official method FO-13, Determination of Sulphated Ash in Citrus Red No. 2;

"official method FO-14" means official method FO-14, Determination of Water Soluble Matter in Citrus Red No. 2;

"official method FO-15" means official method FO-15, Determination of Carbon Tetrachloride Insoluble Matter in Citrus Red No. 2;

"preparation" means a preparation of one or more synthetic colours containing less than three per cent dye and sold for household use;

"synthetic colour" means any organic **colour,** other than caramel, **that is produced by chemical synthesis and has no counterpart in nature** and for which a standard is prescribed **in section B.06.041 to B.06.053.**

B.06.002.

No person shall sell a food, other than a synthetic colour, mixture, preparation or flavouring preparation, **that contains,** when prepared for consumption according to label directions, **more than**

(a) **300 parts per million of Allura Red, Amaranth, Erythrosine, Tartrazine, Sunset Yellow FCF or Indigotine or any combination of these colours;**
(b) **100 parts per million of Fast Green FCF or Brilliant Blue FCF or any combination** of those colours;
(c) **300 parts per million of any combination of the synthetic colours named in paragraphs (a) and (b)** within the limits set by those paragraphs; **or**
(d) **150 parts per million of Ponceau SX**

B.06.003.

No person shall sell a colour for use in or upon food that contains more than

(a) **three parts per million of arsenic,** calculated as arsenic, as determined by official method FO-4,
(b) **10 parts per million of lead, calculated as lead,** as determined by official method FO-5, or
(c) **40 parts per million of heavy metals,** except in the case of iron oxide, titanium dioxide, aluminum metal and silver metal.

B.06.005.
No person shall import a synthetic colour for use in or upon food **unless**

(a) the Director, or an agency acceptable to the Director, has certified that each lot meets the requirements of section B.06.003 and the standard for such colour as prescribed in sections B.06.041 to B.06.053; and
(b) where **the synthetic colour is certified by an agency,** a copy of the certificate has been submitted to **and accepted by the Director.**

B.06.008.
No person shall import or **sell a mixture or preparation for use in or upon food** unless

(a) **the Director,** or an agency acceptable to the Director, **has certified that any synthetic colour** contained therein **meets the requirements of section B.06.003** and the standard for such colour as prescribed in section B.06.041 to B.06.053; or
(b) any synthetic colour contained therein has been previously certified and the certificate has been accepted as required by sections B.06.004 and B.06.005.

Natural Colours

B.06.021. [S].
Oil-soluble Annatto, Annatto Butter Colour, or Annatto Margarine Colour

(a) shall be the extractives of *Bixa orellana* seeds
  (i) dissolved in vegetable oil, caster oil, **monoglycerides** and **diglycerides, propylene glycol** or **propylene glycol monoesters and diesters of fat-forming fatty acids** with or without 1.0 per cent potassium hydroxide, or
  (ii) suspended upon **sugar, lactose, starch or hydrated calcium silicate, with or without calcium phosphate, potassium aluminum sulphate, sodium bicarbonate or salt;** and
(b) shall contain not less than 0.30 per cent total pigments calculated as natural (cis-) bixin, which pigments shall consist of not less than 40 per cent natural (cis-) bixin.

B.06.022. [S].
ß-Carotene shall be the food colour chemically known as ß-carotene that is manufactured synthetically and shall conform to the following specifications:

(a) **1 per cent solution in chloroform** .... clear;
(b) loss of weight on drying .... not more than 0.2 per cent;
(c) residue on ignition .... not more than 0.2 per cent; and
(d) assay (spectrophotometric) .... 96 to 101 per cent.

B.06.023. [S].
ß-Apo-8'-Carotenal shall be the food colour chemically known as ß-apo-8'-carotenal and shall conform to the following specifications:

(a) **1 per cent solution in chloroform** .... clear;
(b) melting point (decomposition) ..... 136°C to 140°C (corrected);
(c) loss of weight on drying .... not more than 0.2 per cent;
(d) residue on ignition .... not more than 0.2 per cent; and
(e) assay (spectrophotometric) .... 96 to 101 per cent.

B.06.024. [S].
Canthaxanthin shall be the food colour chemically known as canthaxanthin and shall conform to the following specifications:

(a) **1 per cent solution in chloroform** .... clear;
(b) loss of weight on drying. .... not more than 0.2 per cent; and
(c) assay (spectrophotometric) .... 96 to 101 per cent.

B.06.025. [S].
Ethyl ß-apo-8'-carotenoate shall be the food colour chemically known as ethyl ß-apo-8'-carotenoate and shall conform to the following specifications:

(a) **1 per cent solution in chloroform**. .... clear;
(b) loss of weight on drying .... not more than 0.2 per cent; and
(c) assay (spectrophotometric) .... 96 to 101 per cent.

Inorganic Colours

B.06.031. [S].
Carbon black shall be carbon prepared from natural gas by the "channel" or "impingement" process **and shall contain no higher aromatic hydrocarbons or tarry materials** as determined by official method FO-6, Determination of Higher Aromatic Hydrocarbons or Tarry Materials in Carbon Black and Charcoal.

B.06.032. [S].
Charcoal shall be carbon prepared by the incomplete combustion of vegetable matter and shall contain no higher aromatic hydrocarbons or tarry materials as determined by official method FO-6.

B.06.033 [S].
Titanium Dioxide shall be the chemical substance **known as titanium dioxide** and shall contain not less than 99 per cent titanium dioxide, and notwithstanding section B.06.006, shall contain not more than 50 parts per million of total antimony expressed as the metal and as determined by an acceptable method.

Synthetic Colours

B.06.041. [S].
Amaranth shall be the trisodium salt of 1-(4-sulpho-1-naphthylazo)-2-naphthol-3,6-disulphonic acid, shall contain not less than 85 per cent dye as determined by official method FO-7, and may contain not more than

(a) 0.2 per cent water insoluble matter, as determined by official method FO-8;
(b) 0.2 per cent combined ether extracts, as determined by official method FO-9;
(c) 4.0 per cent subsidiary dyes, as determined by official method FO-10; and
(d) 0.5 per cent intermediates, as determined by official method FO-11.

B.06.042. [S].
Erythrosine shall be the disodium salt of 2,4,5,7-tetraiodofluorescein, shall contain not less than 85 per cent dye calculated as the monohydrate, as determined by official method FO-7 and may contain not more than

(a) 0.2 per cent water insoluble matter, as determined by official method FO-8;
(b) 0.2 per cent combined ether extracts, as determined by official method FO-9;
(c) 5.0 per cent subsidiary dyes, as determined by official method FO-10; and
(d) 0.5 per cent intermediates, as determined by official method FO-11.

B.06.043. [S].
Ponceau SX shall be the disodium salt of 2-(5-sulpho-2,4-xylylazo)-1-naphthol-4-sulphonic acid, shall contain not less than 85 per cent dye, as determined by official method FO-7 and may contain not more than

(a) 0.2 per cent water insoluble matter, as determined by official method FO-8;
(b) 0.2 per cent combined ether extracts, as determined by official method FO-9;
(c) 1.0 per cent subsidiary dyes, as determined by official method FO-10; and
(d) 0.5 per cent intermediates, as determined by official method FO-11.

B.06.044. [S]. Allura Red shall be the disodium salt of 6-hydroxy-5-[(2-methoxy-5-methyl-4-sulphophenyl)azo]-2-naphthalene sulphonic acid, shall contain not less than 85 per cent dye, as determined by official method FO-7 and may contain not more than

(a) 0.2 per cent water insoluble matter, as determined by official method FO-8;
(b) 0.2 per cent combined ether extracts, as determined by official method FO-9;
(c) 4.0 per cent subsidiary dyes, as determined by official method FO-10; and
(d) 0.5 per cent intermediates, as determined by official method FO-11.

B.06.045. [S].
Tartrazine shall be the trisodium salt of 3-carboxy-5-hydroxy-1-*p*-sulphophenyl-4-*p*-sulpho-phenylazopyrazole, shall contain not less than 85 per cent dye, as determined by official method FO-7, and may contain not more than

(a) 0.2 per cent water insoluble matter, as determined by official method FO-8;
(b) 0.2 per cent combined ether extracts, as determined by official method FO-9;
(c) 1.0 per cent subsidiary dyes, as determined by official method FO-10; and
(d) 0.5 per cent intermediates, as determined by official method FO-11.

B.06.046. [S].
Sunset Yellow FCF shall be the disodium salt of 1-*p*-sulphophenylazo-2-naphthol-6-sulphonic acid, shall contain not less than 85 per cent dye, as determined by official method FO-7 and may contain not more than
- (a) 0.2 per cent water insoluble matter, as determined by official method FO-8;
- (b) 0.2 per cent combined ether extracts, as determined by official method FO-9;
- (c) 5.0 per cent subsidiary dyes, as determined by official method FO-10; and
- (d) 0.5 per cent intermediates, as determined by official method FO-11.

B.06.049. [S].
Fast Green FCF shall be the disodium salt of 4,4'-di (*N*-ethyl-*m*-sulphobenzyl-amino)-2"-sulpho-4"-hydroxytriphenylmethanol anhydride, shall contain not less than 85 per cent dye, as determined by official method FO-7 and may contain not more than
- (a) 0.2 per cent water insoluble matter, as determined by official method FO-8;
- (b) 0.4 per cent combined ether extracts, as determined by official method FO-9;
- (c) 5.0 per cent subsidiary dyes, as determined by official method FO-10; and
- (d) 1.0 per cent intermediates, as determined by official method FO-11.

B.06.050. [S].
Indigotine shall be the disodium salt of indigotine-5,5'-disulphonic acid, shall contain not less than 85 per cent dye, as determined by official method FO-7 and may contain not more than
- (a) 0.2 per cent water insoluble matter, as determined by official method FO-8;
- (b) 0.4 per cent combined ether extracts, as determined by official method FO-9;
- (c) 1.0 per cent subsidiary dyes, as determined by official method FO-10; and
- (d) 0.5 per cent intermediates, as determined by official method FO-11.

B.06.051. [S].
Brilliant Blue FCF shall be the disodium salt of 4,4'-di (*N*-ethyl-*m*-sulphobenzyl-amino)-2"-sulphotriphenylmethanol anhydride, shall contain not less than 85 per cent dye, as determined by official method FO-7 and may contain not more than

(a) 0.2 per cent water insoluble matter, as determined by official method FO-8;
(b) 0.4 per cent combined ether extracts, as determined by official method FO-9;
(c) 5.0 per cent subsidiary dyes, as determined by official method FO-10; and
(d) 2.0 per cent intermediates, as determined by official method FO-11.

B.06.053. [S].
Citrus Red No. 2 shall be 1-(2,5-dimethoxyphenylazo)-2-naphthol, shall contain not less than 98 per cent dye, as determined by official method FO-7 and may contain not more than

(a) 0.5 per cent volatile matter (at 100°C), as determined by official method FO-12;
(b) 0.3 per cent sulphated ash, as determined by official method FO-13;
(c) 0.3 per cent water soluble matter, as determined by official method FO-14;
(d) 0.5 per cent carbon tetrachloride insoluble matter, as determined by official method FO-15;
(e) 0.05 per cent uncombined intermediates; and
(f) 2.0 per cent subsidiary dyes.

B.06.061.
The lake of any water soluble synthetic colour for which a standard is provided in sections B.06.041, B.06.042, B.06.043, B.06.044, B.06.045, B.06.046, B.06.049, B.06.050 or B.06.051 **shall be the calcium or aluminum salt of the respective colour extended on alumina.**

DIVISION 7

Spices, Dressings and Seasonings

B.07.003. [S]. Basil or **Sweet Basil,** whole or ground, shall be the dried leaves of *Ocimum basilicum L.* and shall contain

(a) not more than
(i) 15 per cent total ash,
(ii) 2 per cent ash insoluble in hydrochloric acid, and
(iii) 9 per cent moisture; and
(b) not less than 0.2 millilitre volatile oil per 100 grams of spice.

B.07.004. [S].
Bay leaves or **Laurel Leaves,** whole or ground, shall be the dried leaves of *Laurus nobilis L.* and shall contain

(a) not more than
(i) 4.5 per cent total ash,
(ii) 0.5 per cent ash insoluble in hydrochloric acid, and
(iii) 7 per cent moisture; and
(b) not less than 1 millilitre volatile oil per 100 grams of spice.

B.07.007. [S].
Cayenne Pepper or **Cayenne,** whole or ground,

(a) shall be the dried ripe fruit of *Capsicum frutescens L.*, *Capsicum baccatum L.*, or other small-fruited species of Capsicum and shall contain not more than
(i) 1.5 per cent starch,
(ii) 28 per cent crude fibre,
(iii) 10 per cent total ash,
(iv) 1.5 per cent ash insoluble in hydrochloric acid, and
(v) 10 per cent moisture; and
(b) may contain **silicon dioxide as an anticaking agent** in an amount not exceeding 2.0 per cent.

B.07.008. [S].
Celery Salt

(a) shall be a combination of
(i) ground celery seed or ground dehydrated celery, and
(ii) salt in an amount not exceeding 75 per cent; and
(b) may contain **silicon dioxide as an anticaking agent** in an amount not exceeding 0.5 per cent.

B.07.015. [S].
Cumin or **Cumin Seed,** whole or ground, shall be the dried seeds of *Cuminum cyminum L.* and shall contain
(a) not more than
(i) 9.5 per cent total ash
(ii) 1.5 per cent ash insoluble in hydrochloric acid, and
(iii) 9 per cent moisture; and
(b) not less than 2.5 millilitres volatile oil per 100 grams of spice.

B.07.016. [S].
Curry Powder shall be any combination of
(a) turmeric with spices and seasoning; and
(b) salt in an amount not exceeding 5 per cent.

B.07.020. [S]. Garlic Salt
(a) shall be a combination of
(i) powdered dehydrated garlic, and
(ii) salt in an amount not exceeding 75 per cent; and
(b) may contain one or more of the following **anticaking agents** in a total amount not exceeding 2 per cent: **calcium aluminum silicate, calcium phosphate tribasic, calcium silicate, calcium stearate, magnesium carbonate, magnesium silicate, magnesium stearate, silicone dioxide** in an amount not exceeding 1 per cent and sodium aluminum silicate.

B.07.025. [S].
Mustard, Mustard Flour or **Ground Mustard** shall be powdered mustard seed
(a) made from mustard seed from which
(i) most of the hulls have been removed, and
(ii) a portion of the fixed oil may have been removed; and
(b) that contains not more than
(i) 1.5 per cent starch, and
(ii) 8 per cent total ash, on an oil-free basis.

B.07.027. [S].
Onion Salt
(a) shall be a combination of
(i) powdered dehydrated onion, and
(ii) salt in an amount not exceeding 75 per cent; and

(b) may contain one or more of the following **anticaking agents** in a total amount not exceeding 2 per cent: **calcium aluminum silicate, calcium phosphate tribasic, calcium silicate, calcium stearate, magnesium carbonate, magnesium silicate, magnesium stearate, silicone dioxide** in an amount not exceeding 1 per cent and sodium aluminum silicate.

B.07.028. [S].
Oregano, whole or ground, shall be the dried leaves of *Origanum vulgar L.* or *Origanum Spp.* and shall contain
- (a) not more than
  - (i) 10 per cent total ash,
  - (ii) 2 per cent ash insoluble in hydrochloric acid, and
  - (iii) 10 per cent moisture; and
- (b) not less than 2.5 millilitres volatile oil per 100 grams of spice.

B.07.029. [S]. Paprika shall be the dried, ground ripe fruit of *Capsicum annuum L.* and
- (a) shall contain not more than
  - (i) 23 per cent crude fibre;
  - (ii) 8.5 per cent total ash,
  - (iii) 1 per cent ash insoluble in hydrochloric acid, and
  - (iv) 12 per cent moisture; and
- (b) may contain **silicon dioxide as an anticaking agent** in an amount not exceeding 2.0 per cent.

B.07.040. [S].
Mayonnaise, Mayonnaise Dressing or Mayonnaise Salad Dressing
- (a) shall be a combination of
  - (i) vegetable oil
  - (ii) **whole egg or egg yolk, in liquid, frozen or dried form** and
  - (iii) vinegar or lemon juice;
- (b) may contain
  - (i) water,
  - (ii) salt,
  - (iii) a sweetening agent,
  - (iv) spice or other seasoning except turmeric or saffron,
  - (v) citric, tartaric or lactic acid, and
  - (vi) **a sequestering agent;** and
- (c) shall contain not less than 65 per cent vegetable oil.

B.07.041. [S].
French Dressing

(a) shall be a combination of
  (i) vegetable oil, and
  (ii) vinegar or lemon juice;
(b) may contain
  (i) water,
  (ii) salt,
  (iii) a sweetening agent,
  (iv) spice, tomato or other seasoning,
  (v) **an emulsifying agent,**
  (vi) **whole egg or egg yolk, in liquid, frozen or dried form,**
  (vii) citric, tartaric or lactic acid, and
  (viii) **a sequestering agent;** and
(c) shall contain not less than 35 per cent vegetable oil.

B.07.042. [S].
Salad Dressing

(a) shall be a combination of
  (i) vegetable oil,
  (ii) whole egg or egg yolk, in liquid, frozen or dried form,
  (iii) vinegar or lemon juice, and
  (iv) starch, flour, rye flour or tapioca flour or any combination thereof;
(b) may contain
  (i) water,
  (ii) salt,
  (iii) a sweetening agent,
  (iv) spice or other seasoning,
  (v) **an emulsifying agent,**
  (vi) citric, tartaric or lactic acid, and
  (vii) **a sequestering agent;** and
(c) shall contain not less than 35 per cent vegetable oil.

DIVISION 8

Dairy Products

B.08.001.
The foods referred to in this division are dairy products.

Authors' Note: *Recombinant bovine growth hormone (rBGH), also known as recombinant bovine somatotrophin, is a genetically engineered hormone injected into dairy cows to increase their milk production.*

At first, researchers were concerned about the safety of rBGH because it increased the incidence of birth defects in cows. It also increased the rate of udder infection which, in turn, caused pus and antibiotics to appear in the milk. Naturally, this posed a risk to humans who drank the milk. Currently, researchers are warning the public to beware of products containing rBGH as there is a possibility that people who drink the milk from hormone treated cows will stand a greater chance of developing breast, colon or prostate cancer. Researchers claim that the milk from these cows differs from natural milk in its chemical, nutritioinal and pharmacological makeup. This research is detailed in prestigious medical journals such as The Lancet *and the* International Journal of Health Services.

It is approved for use in the United States. As of the printing of this book, Canadian law forbids the use of rBGH in dairy cows, but Monsanto Canada, the drug's manufacturer, will be appealing this regulation. Keep your eyes on the news – and your food.

B.08.003. [S].
Milk or **Whole Milk**

(a) shall be the normal lacteal secretion obtained from the mammary gland of the cow, genus *Bos*; and
(b) shall contain added vitamin D in such an amount that a reasonable daily intake of the milk contains not less than 300 International Units and not more than 400 International Units of vitamin D.

B.08.004. [S].
Skim Milk

(a) shall be milk that contains not more than 0.3 per cent milk fat;
(b) shall, notwithstanding sections D.01.009 and D.01.010, contain added vitamin A in such an amount that a reasonable daily intake

of the milk contains not less than 1,200 International Units and not more than 2,500 International Units of vitamin A; and

(c) shall contain vitamin D in such an amount that a reasonable daily intake of the milk contains not less than 300 International Units and not more than 400 International Units of vitamin D.

B.08.009. [S].

Condensed Milk or **Sweetened Condensed Milk** shall be milk from which water has been evaporated and to which has been added sugar, dextrose, glucose, glucose solids or lactose, or any combination thereof, may contain added vitamin D and shall contain not less than

(a) 28 per cent milk solids, and
(b) 8 per cent milk fat.

B.08.010. [S]. Evaporated Milk

(a) shall be milk from which water has been evaporated;
(b) shall contain not less than
    (i) 25.0 per cent milk solids, and
    (ii) 7.5 per cent milk fat;
(c) shall, notwithstanding sections D.01.009 to D.01.011, contain added vitamin C in such an amount that a reasonable daily intake of the milk contains not less than 60 milligrams and not more than 75 milligrams of vitamin C;
(d) shall contain vitamin D in such an amount that a reasonable daily intake of the milk contains not less than 300 International Units and not more than 400 International Units of vitamin D; and
(e) may contain
    (i) added **disodium phosphate or sodium citrate, or both,** and
    (ii) **an emulsifying agent.**

B.08.013. [S].

Milk Powder, Whole Milk Powder, Dry Whole Milk, or **Powdered Whole Milk**

(a) shall be dried milk;
(b) shall contain not less than
    (i) 95 per cent milk solids, and
    (ii) 26 per cent milk fat;

(c) shall contain added vitamin D in such an amount that a reasonable daily intake of the milk contains not less than 300 International Units and not more than 400 International Units of vitamin D; and
(d) may contain the emulsifying agent lecithin in an amount not exceeding 0.5%.

B.08.015.
No person shall sell milk, skim milk, partly skimmed milk, (naming the flavour) milk, (naming the flavour) skim milk, (naming the flavour) partly skimmed milk, skim milk with added milk solids, partly skimmed milk with added milk solids, (naming the flavour) skim milk with added milk solids, (naming the flavour) partly skimmed milk with added milk solids, condensed milk, evaporated milk, evaporated skim milk, evaporated partly skimmed milk, milk powder or skim milk powder, **in which the vitamin content has been increased by either irradiation or addition unless**
(a) in the case of the addition of vitamin D, the menstruum containing the vitamin D contributes not more than 0.01 per cent fat foreign to milk;
(b) in cases where the **vitamin D content is increased by irradiation, the principal display panel of the label carries the statement "Vitamin D Increased"** immediately preceding or following the name of the food without intervening written, printed or graphic matter.

B.08.016. [S].
(naming the flavour) Milk
(a) shall be the product made from
  (i) milk, milk powder, skim milk, skim milk powder, or partly skimmed milk, evaporated milk, evaporated partly skimmed milk, evaporated skim milk or cream or any combination thereof,
  (ii) **a flavouring preparation,** and
  (iii) a sweetening agent;
(b) shall contain not less than 3.0 per cent milk fat;
(c) shall contain added vitamin D in such an amount that a reasonable daily intake of the milk contains not less than 300 International Units and not more than 400 International Units of vitamin D;

(d) **may contain salt, food colour, lactase, stabilizing agent** and not more than 0.5 per cent **starch**; and

(e) may contain not more than 50,000 total aerobic bacteria per cubic centimetre, as determined by official method MFO-7, Microbiological Examination of Milk.

B.08.021. [S].
Malted Milk or **Malted Milk Powder**

(a) shall be the product made by combining milk with the liquid separated from a mash of ground barley malt and meal;

(b) may have added to it, in such manner as to secure the full enzyme action of the malt extract, salt and sodium bicarbonate or potassium bicarbonate;

(c) may have water removed from it; and

(d) shall contain

  (i) not less than 7.5 per cent milk fat, and

  (ii) not more than 3.5 per cent moisture.

B.08.023. [S].
(naming the flavour) Skim Milk with Added Milk Solids

(a) shall be the product made from

  (i) skim milk, skim milk powder, or evaporated skim milk or any combination thereof,

  (ii) **a flavouring preparation,** and

  (iii) a sweetening agent;

(b) shall contain not less than 10 per cent milk solids not including fat;

(c) shall contain not more than 0.3 per cent milk fat;

(d) shall, notwithstanding sections D.01.009 and D.01.010, contain added vitamin A in such an amount that a reasonable daily intake of the milk contains not less than 1,200 International Units and not more than 2,500 International Units of vitamin A;

(e) shall contain added vitamin D in such an amount that a reasonable daily intake of the milk contains not less than 300 International Units and not more than 400 International Units of vitamin D; and

(f) **may contain salt, food colour, lactase, stabilizing agent** and not more than 0.5 per cent **starch.**

B.08.027.
Notwithstanding anything contained in these Regulations, the following dairy products that are used in or sold for the manufacture of other foods are not required to contain added vitamins: milk, partly skimmed milk, partially skimmed milk, skim milk, sterilized milk, evaporated milk, evaporated skim milk, concentrated skim milk, evaporated partly skimmed milk, concentrated partly skimmed milk, milk powder, dry whole milk, powdered whole milk, skim milk powder, dry skim milk, partly skimmed milk powder, partially skimmed milk powder, skim milk with added milk solids, partly skimmed milk with added milk solids, and partially skimmed milk with added milk solids.

Goat's Milk

B.08.028.1 A lacteal secretion obtained from the mammary gland of any animal other than a cow, genus *Bos*, and a product or derivative of such secretion shall be labelled so as to identify that animal.

B.08.029.
(1) Notwithstanding sections D.01.009 to D.01.011, no person shall sell goat's milk or goat's milk powder to which vitamin D has been added unless the goat's milk or goat's milk powder contains not less than 35 International Units and not more than 45 International Units of vitamin D per 100 mL of the food when ready-to-serve.

Authors' Note: *Among the many additives permitted in cheese (listed below), the addition of artificial flavouring and colouring is the most unfortunate. Such additives are extremely dangerous and one can only question their necessity in what is otherwise a nutritious food. Garishly coloured orange cheese does not taste better than its natural cream-coloured counterpart.*

Because labelling can be so misleading, many consumers may not realize the number of dangerous additives that can be used in the manufacture of cheese. The following excerpt demonstrates that potassium nitrate and sodium nitrate, among many other chemicals, may be added to cheese.

B.08.033 (1) [S].
(Naming the variety) Cheese, other than cheddar cheese, cream cheese, whey cheese, cream cheese with (named added ingredients), cream cheese spread, cream cheese spread with (named added

ingredients), processed (named variety) cheese, processed (named variety) cheese with (named added ingredients), processed cheese food, processed cheese food with (named added ingredients), processed cheese spread, processed cheese spread with (named added ingredients), cold-pack (named variety) cheese, cold-pack (named variety) cheese with (named added ingredients), cold-pack cheese food, cold-pack cheese food with (named added ingredients), cottage cheese and creamed cottage cheese,

(a) shall
  - (i) be the product made by coagulating milk, milk products or a combination thereof with the aid of bacteria to form a curd and forming the curd into a homogeneous mass after draining the whey,
  - (ii) possess the physical, chemical and organoleptic properties typical for the variety,
  - (iii) where it is a cheese of a variety named in the Table to this section, contain no more than the maximum percentage of moisture shown in column II thereof for that variety,
  - (iv) where it is a cheese of a variety named in Part I of the Table to this section, contain no less than the minimum percentage of milk fat shown in column III for that variety, and
  - (v) where it is cheese of a variety named in Part II of the Table to this section, contain no more than the maximum percentage of milk fat shown in column III for that variety; and

(b) may contain
  - (i) salt, seasonings, condiments and **spices,**
  - (ii) **flavouring preparations** other than cheese flavouring,
  - (iii) **micro-organisms** to aid further ripening,
  - (iv) **one or more of the following colouring agents:**
    - (A) in an amount consistent with good manufacturing practice, annatto, beta-carotene, chlorophyll, paprika, riboflavin or turmeric, and
    - (B) in an amount not exceeding 35 parts per million, beta-apo-8'-carotenal, ethyl beta-apo-8'-carotenoate or a combination thereof,
  - (v) **calcium chloride as a firming agent** in an amount not exceeding 0.02 per cent of the milk and milk products used,

(vi) **paraffin wax as a coating** in an amount consistent with good manufacturing practice,
(vii) **where potassium nitrate, sodium nitrate or a combination thereof** are used for the purpose and in the manner described in subsection (2), residues of potassium nitrate, sodium nitrate or a combination thereof in an amount not exceeding 50 parts per million,
(viii) **wood smoke as a preservative** in an amount consistent with good manufacturing practice, **and**
(ix) **the following preservatives:**
(A) **propionic acid, calcium propionate, sodium propionate or any combination thereof** in an amount not exceeding 2,000 parts per million, calculated as propionic acid,
(B) **sorbic acid, calcium sorbate, potassium sorbate, sodium sorbate, or any combination thereof** in an amount not exceeding 3,000 parts per million, calculated as sorbic acid,
(C) **any combination of the preservatives named in clauses (A) and (B)** in an amount not exceeding 3,000 parts per million, calculated as propionic acid and sorbic acid respectively, or
(D) **natamycin** applied to the surface of the cheese, in an amount not exceeding 20 parts per million, calculated on the weight of the cheese, and
(x) **in the case of grated cheese, calcium silicate, microcrystalline cellulose or cellulose, or a combination thereof, as an anticaking agent,** the total amount not to exceed 2.0 per cent.

(2) **Potassium nitrate, sodium nitrate or a combination thereof may be used as a preservative in cheese providing the following requirements are met:**
(a) the amount of the salt or combination of salts does not exceed 200 parts per million of the milk and milk products used to make the cheese;
(b) the cheese in which **the preservative is** used is
(i) mold ripened cheese packed in a hermetically sealed container,
(ii) ripened cheese

(A) that contains not more than 68 percent moisture on a fat free basis, and
(B) during the manufacture of which the lactic acid fermentation and salting was completed more than 12 hours after coagulation of the curd by enzymes; and
(c) **the salting is,** in the case of the cheese **described in subparagraph (b)(ii),** applied externally **as a dry salt or in the form of a brine.**

(3) No person shall use an enzyme other than
(a) a milk coagulating enzyme derived from *Mucor miehei* (Cooney and Emmerson) or *Mucor pusillus Lindt*, **Chymosin A** derived from *Escherichia coli* K-12, **GE81 (pPFZ87A), Chymosin B** derived from *Aspergillus niger* var. *awamori*, GCC0349 **(pGAMpR)** or from *Kluyveromyces marxianus* var. *lactis*, **DS1182 (pKS105)**, protease derived from *Micrococcus caseolyticus*, pepsin, rennet or bovine rennet, in the manufacture of any cheese to which subsection (1) applies;
(b) lipase and enzymes described in paragraph (a), in the manufacture of Asiago cheese, Blue cheese, Caciocavallo cheese, Feta cheese and Provolone cheese;
(c) a milk coagulating enzyme derived from *Endothia parasitica* and enzymes described in paragraph (a), in the manufacture of Emmentaler (Emmental, Swiss) cheese, Mozzarella (Scamorza) cheese and Part Skim Mozzarella (Part Skim Scamorza) cheese;
(d) lipase, a milk coagulating enzyme derived from *Endothia parasitica* and enzymes described in paragraph (a), in the manufacture of Parmesan cheese and Romano cheese; and
(e) protease derived from *Aspergillus oryzae*, in the manufacture of Colby cheese.

B.08.034 (1) [S].
Cheddar Cheese
(a) shall
(i) be the product made by coagulating milk, milk products or a combination thereof with the aid of bacteria to form a curd and subjecting the curd to
(A) the cheddar process, or
(B) any process other than the cheddar process that produces a cheese having the same physical,

chemical and organoleptic properties as those of cheese produced by the cheddar process, and

(ii) contain

(A) not more than 39 per cent moisture, and

(B) not less than 31 per cent milk fat; and

(b) may contain

(i) salt,

(ii) **flavouring preparations** other than cheese flavouring,

(iii) bacterial cultures to aid further ripening,

(iv) one or more of the following colouring agents:

(A) in an amount consistent with good manufacturing practice, annatto, beta-carotene, chlorophyll, paprika, riboflavin, turmeric, and

(B) in an amount not exceeding 35 parts per million, either singly or in combination thereof, beta-apo-8'-carotenal, ethyl beta-apo-8'-carotenoate,

(v) **calcium chloride as a firming agent** in an amount not exceeding 0.02 per cent of the milk and milk products used,

(vi) **wood smoke as a preservative** in an amount consistent with good manufacturing practice,

(vii) **the following preservatives:**

(A) **propionic acid, calcium propionate, sodium propionate or a combination thereof** in an amount not exceeding 2,000 parts per million calculated as propionic acid,

(B) **sorbic acid, calcium sorbate, potassium sorbate, sodium sorbate or a combination thereof** in an amount not exceeding 3,000 parts per million calculated as sorbic acid,

(C) **any combination of the preservatives named in clauses (A) and (B)** in an amount not exceeding 3,000 parts per million calculated as propionic acid and sorbic acid respectively, or

(D) **natamycin** applied to the surface of the cheese, in an amount not exceeding 20 parts per million, calculated on the weight of the cheese, and

(viii) **in the case of grated cheddar cheese, calcium silicate, microcrystalline cellulose or cellulose, or a**

combination thereof, as an anticaking agent, the total amount not to exceed 2.0 per cent.

(2) No person shall, in the manufacture of the cheddar cheese, use any enzyme other than
(a) lipase, pepsin, rennet, milk coagulating enzyme derived from *Mucor miehei* (Cooney and Emmerson) or *Mucor pusillus Lindt*, **Chymosin A** derived from *Escherichia coli* **K-12, GE81 (pPFZ87A), Chymosin B** derived from *Aspergillus niger* var. *awamori*, **GCC0349 (pGAMpR)** or from *Kluyveromyces marxianus* var. *lactis*, **DS1182 (pKS105);** and
(b) protease derived from *Aspergillus oryzae*.

(2.1) **No person shall use an enzyme** referred to in subsection (2) **at a level of use above that consistent with good manufacturing practice.**

(3) Where a flavouring preparation is added to a cheese as permitted in subsection (1), the words "with (naming the flavouring preparation)" shall be added to the common name in any label.

(4) Only a cheese to which wood smoke has been added as permitted in subsection (1) may be described by the term "smoked" on a label.

(5) Where a cheese is labelled as permitted in subsection (4), the word "smoked" shall be shown on the principal display panel.

B.08.035. (1) [S].
Cream Cheese
(a) shall
(i) be the product made by coagulating cream with the aid of bacteria to form a curd and forming the curd into a homogeneous mass after draining the whey, and
(ii) contain
(A) not more than 55 per cent moisture, and
(B) not less than 30 per cent milk fat; and
(b) may contain
(i) cream added to adjust the milk fat content,
(ii) salt,

(iii) nitrogen to improve spreadability in an amount consistent with good manufacturing practice,

(iv) **the following emulsifying, gelling, stabilizing and thickening agents: ammonium carrageenan, calcium carrageenan, carob bean gum (locust bean gum), carrageenan, gelatin, guar gum, Irish Moss Gelose, potassium carrageenan, propylene glycol alginate, sodium carboxymethyl cellulose (carboxymethyl cellulose, cellulose gum, sodium cellulose glycolate), sodium carrageenan, tragacanth gum, xanthan gum or any combination thereof** in an amount not exceeding 0.5 per cent, and

(v) **the following preservatives:**

A) **propionic acid, calcium propionate, sodium propionate or any combination thereof** in an amount not exceeding 2,000 parts per million, calculated as propionic acid,

(B) **sorbic acid, calcium sorbate, potassium sorbate, sodium sorbate or any combination thereof** in an amount not exceeding 3,000 parts per million, calculated as sorbic acid, or

(C) **any combination of the preservatives named in clauses (A) and (B)** in an amount not exceeding 3,000 parts per million, calculated as propionic acid and sorbic acid respectively.

(2) **No person shall use any enzyme**

(a) other than **Chymosin A** derived from *Escherichia coli* K-12, **GE81 (pPFZ87A), Chymosin B** derived from *Aspergillus niger* var. *awamori*, **GCC0349 (pGAMpR)** or from *Kluyveromyces marxianus* var. *lactis*, **DS1182 (pKS105),** pepsin or rennet in the manufacture of a product to which subsection (1) applies; and

(b) **at a level of use above that consistent with good manufacturing practice.**

B.08.038. (1) [S].
Cream Cheese Spread

(a) shall

(i) be the product made by coagulating cream with the aid

of bacteria to form a curd and forming the curd into a homogeneous mass after draining the whey, and

(ii) contain

(A) added milk and milk products,

(B) not less than 51 per cent cream cheese,

(C) not more than 60 per cent moisture, and

(D) not less than 24 per cent milk fat; and

(b) may contain

(i) cream added to adjust the milk fat content,

(ii) salt, vinegar and sweetening agents,

(iii) nitrogen to improve spreadability in an amount consistent with good manufacturing practice,

(iv) **one or more of the following colouring agents:**

(A) in an amount consistent with good manufacturing practice, annatto, beta-carotene, chlorophyll, paprika, riboflavin, turmeric, and

(B) in an amount not exceeding 35 parts per million, either singly or in combination thereof, beta-apo-8'-carotenal, ethyl beta-apo-8'-carotenoate,

(v) **the following emulsifying, gelling, stabilizing and thickening agents:**

(A) **ammonium carrageenan, calcium carrageenan, carob bean gum (locust bean gum), carrageenan, gelatin, guar gum, Irish Moss Gelose, potassium carrageenan, propylene glycol alginate, sodium carboxymethyl cellulose (carboxymethyl cellulose, cellulose gum, sodium cellulose glycolate), sodium carrageenan, tragacanth gum, xanthan gum or a combination thereof** in an amount not exceeding 0.5 per cent, and

(B) **calcium phosphate dibasic, potassium phosphate dibasic, sodium acid pyrophosphate, sodium aluminum phosphate, sodium hexametaphosphate, sodium phosphate dibasic, sodium phosphate monobasic, sodium phosphate tribasic, sodium pyrophosphate tetrabasic, calcium citrate, potassium citrate, sodium citrate, sodium potassium tartrate, sodium tartrate, sodium gluconate or any**

combination thereof in an amount when calculated as anhydrous salts, not exceeding 3.5 per cent in the case of phosphate salts and 4.0 per cent in total,

(vi) **acetic acid, calcium carbonate, citric acid, lactic acid, malic acid, phosphoric acid, potassium bicarbonate, potassium carbonate, sodium bicarbonate, sodium carbonate and tartaric acid as pH-adjusting agents in an amount consistent with good manufacturing practice, and**

(vii) **the following preservatives:**

(A) **propionic acid, calcium propionate, sodium propionate or any combination thereof** in an amount not exceeding 2,000 parts per million, calculated as propionic acid,

(B) **sorbic acid, calcium sorbate, potassium sorbate, sodium sorbate, or any combination thereof** in an amount not exceeding 3,000 parts per million, calculated as sorbic acid, or

(C) **any combination of the preservatives named in clauses (A) and (B)** in an amount not exceeding 3,000 parts per million, calculated as propionic acid and sorbic acid respectively.

(2) **No person shall use any enzyme**

(a) other than **Chymosin A** derived from *Escherichia coli* K-12, **GE81 (pPFZ87A), Chymosin B** derived from *Aspergillus niger* var. *awamori*, **GCC0349 (pGAMpR)** or from *Kluyveromyces marxianus* var. *lactis*, **DS1182 (pKS105)**, pepsin or rennet in the manufacture of a product to which subsection (1) applies; and

(b) at a level of use above that consistent with good manufacturing practice.

Authors' Note: *We have chosen processed cheese as an example of how easily consumers are misled about the actual number of additives contained in certain products. In the excerpt below, you will notice that in the production of processed cheese, the recipe begins with one or more cheeses which are already made up of many additives (see B.08.034). These additives are not required to be listed on the label.*

B.08.040. (1) [S].
Processed (naming the variety) Cheese

(a) shall
- (i) subject to subparagraph (ii), be the product made by comminuting and mixing the named variety or varieties of cheese, other than cream cheese, cottage cheese or whey cheese, into a homogeneous mass with the aid of heat,
- (ii) in the case of processed cheddar cheese, be the product made by comminuting and mixing one or more of the following:
  - (A) cheddar cheese,
  - (B) stirred curd cheese,
  - (C) granular curd cheese, or
  - (D) washed curd cheese into a homogeneous mass with the aid of heat,

(b) **may contain**
- (i) water added to adjust the moisture content,
- (ii) added milk fat,
- (iii) in the case of processed skim milk cheese, added skim milk powder, buttermilk powder and whey powder,
- (iv) salt, vinegar and sweetening agents,
- (v) **one or more of the following colouring agents:**
  - (A) in an amount consistent with good manufacturing practice, annatto, beta-carotene, chlorophyll, paprika, riboflavin, turmeric, and
  - (B) in an amount not exceeding 35 parts per million, either singly or in combination thereof, beta-apo-8'-carotenal, ethyl beta-apo-8'-carotenoate,
- (vi) **the following emulsifying, gelling, stabilizing and thickening agents:**
  - (A) **sodium carboxymethyl cellulose (carboxymethyl cellulose, cellulose gum, sodium cellulose glycolate)** in an amount not exceeding 0.5 per cent,
  - (B) **calcium phosphate dibasic, potassium phosphate dibasic, sodium acid pyrophosphate, sodium aluminum phosphate, sodium hexametaphosphate, sodium phosphate dibasic, sodium phosphate monobasic, sodium phosphate tribasic, sodium pyrophosphate tetrabasic,**

calcium citrate, potassium citrate, sodium citrate, sodium potassium tartrate, sodium tartrate, sodium gluconate or any combination thereof in an amount when calculated as anhydrous salts, not exceeding 3.5 per cent in the case of phosphate salts and 4.0 per cent in total,

(C) lecithin in an amount not exceeding 0.2 per cent, and

(D) **monoglycerides, mono- and diglycerides or any combination thereof** in an amount not exceeding 0.5 per cent,

(vi.1) **calcium phosphate tribasic as an agent to improve colour, texture, consistency and spreadability**, in an amount not exceeding 1 per cent,

(vii) **acetic acid, calcium carbonate, citric acid, lactic acid, malic acid, phosphoric acid, potassium bicarbonate, potassium carbonate, sodium bicarbonate, sodium carbonate and tartaric acid as pH-adjusting agents in an amount consistent with good manufacturing practice,**

(viii) **wood smoke as a preservative** in an amount consistent with good manufacturing practice, and

(ix) **the following preservatives:**

(A) **propionic acid, calcium propionate, sodium propionate or any combination thereof** in an amount not exceeding 2,000 parts per million, calculated as propionic acid,

(B) **sorbic acid, calcium sorbate, potassium sorbate, sodium sorbate, or any combination thereof** in an amount not exceeding 3,000 parts per million, calculated as sorbic acid, or

(C) **any combination of the preservatives named in clauses (A) and (B)** in an amount not exceeding 3,000 parts per million, calculated as propionic acid and sorbic acid respectively.

B.08.041.4. (1) [S].

Processed Cheese Spread with (naming the added ingredients)

(a) shall

(i) be the product made by comminuting and mixing one or more varieties of cheese, other than cream cheese, cottage cheese or whey cheese, into a homogeneous mass with the aid of heat,

(ii) contain the named added ingredients which shall be one or more of the following ingredients in amounts sufficient to differentiate the product from processed cheese spread but not in amounts so large as to change the basic nature of the product:

(A) seasonings, spices, **flavouring preparations,** condiments or chocolate,

(B) fruits, vegetables, pickles, relishes or nuts

(C) prepared or preserved meat, or

(D) prepared or preserved fish, and

(b) **may contain**

(i) water added to adjust the moisture content,

(ii) added milk fat,

(iii) salt, vinegar and sweetening agents,

(iv) **one or more of the following colouring agents:**

(A) in an amount consistent with good manufacturing practice, annatto, beta-carotene, chlorophyll, paprika, riboflavin, turmeric and

(B) in an amount not exceeding 35 parts per million, either singly or in combination thereof, beta-apo-8'-carotenal, ethyl beta-apo-8'-carotenoate,

(v) the following **emulsifying, gelling, stabilizing and thickening agents:**

(A) **ammonium carrageenan, calcium carrageenan, carob bean gum (locust bean gum), carrageenan, gelatin, guar gum, Irish moss gelose, potassium carrageenan, propylene glycol alginate, sodium carboxymethyl cellulose (carboxymethyl cellulose, cellulose gum, sodium cellulose glycolate), sodium carrageenan, tragacanth gum, xanthan gum or a combination thereof** in an amount not exceeding 0.5 per cent,

(B) **calcium phosphate dibasic, potassium phosphate dibasic, sodium acid pyrophosphate, sodium aluminum phosphate, sodium hexametaphosphate, sodium phosphate dibasic, sodium phosphate monobasic, sodium phosphate tribasic, sodium pyrophosphate tetrabasic,**

calcium citrate, potassium citrate, sodium citrate, sodium potassium tartrate, sodium tartrate, sodium gluconate, or any combination thereof in an amount when calculated as anhydrous salts, not exceeding 3.5 per cent in the case of phosphate salts and 4.0 per cent in total,

(C) lecithin in an amount not exceeding 0.2 per cent, and

(D) **monoglycerides, mono- and diglycerides or any combination thereof** in an amount not exceeding 0.5 per cent,

(v.1) **calcium phosphate tribasic as an agent to improve colour, texture, consistency and spreadability,** in an amount not exceeding 1 per cent,

(vi) **acetic acid, calcium carbonate, citric acid, lactic acid, malic acid, phosphoric acid, potassium bicarbonate, potassium carbonate, sodium bicarbonate, sodium carbonate and tartaric acid as pH-adjusting agents in an amount consistent with good manufacturing practice,**

(vii) **wood smoke as a preservative** in an amount consistent with good manufacturing practice, and

(viii) the following **preservatives:**

(A) **propionic acid, calcium propionate, sodium propionate or any combination thereof** in an amount not exceeding 2,000 parts per million, calculated as propionic acid,

(B) **sorbic acid, calcium sorbate, potassium sorbate, sodium sorbate, or any combination thereof** in an amount not exceeding 3,000 parts per million, calculated as sorbic acid, or

(C) **any combination of the preservatives named in clauses (A) and (B)** in an amount not exceeding 3,000 parts per million, calculated as propionic acid and sorbic acid respectively.

B.08.049. [S].
Whey

(a) shall be the product remaining after the curd has been removed from milk in the process of making cheese; and

(b) **may contain**
  (i) **catalase,** in the case of liquid whey **that has been treated with hydrogen peroxide,**
  (ii) lactase,
  (iii) **hydrogen peroxide,** in the case of liquid whey destined for the manufacture of dried whey products, and
  (iv) **benzoyl peroxide and calcium phosphate tribasic,** as a carrier of the benzoyl peroxide, in the case of liquid whey destined for the manufacture of dried whey products other than those for use in infant formula.

B.08.051. [S].
Cottage Cheese
(a) shall be the product, in the form of discrete curd particles, prepared from skim milk, evaporated skim milk or skim milk powder and harmless acid-producing bacterial cultures;
(b) shall contain not more than 80 per cent moisture;
(c) may contain not more than 0.5 per cent stabilizing agent; and
(d) **may contain**
  (i) milk,
  (ii) cream,
  (iii) milk powder
  (iv) **rennet,**
  (v) **milk coagulating enzymes**
  (vi) **Chymosin A**
  (vi.1) **Chymosin B**
  (vii) **salt,**
  (viii) **calcium chloride,**
  (ix) added lactose,
  (x) **pH-adjusting agents,**
  (xi) relishes,
  (xii) fruits, and
  (xiii) vegetables.

B.08.056. [S].
Butter
(a) shall
  (i) be the food prepared in accordance with good manufacturing practices from milk or milk products, and
  (ii) contain not less than 80 per cent milk fat; and

(b) may contain
  (i) milk solids,
  (ii) **bacterial culture,**
  (iii) salt, and
  (iv) **food colour.**

Authors' Note: *The following ice cream products are comprised, in part, of a long list of chemicals including artificial colour and artificial flavour. These excerpts illustrate just how many additives these products contain. In fact, commercial ice cream has become such a smorgasbord of chemical additives that the right to call it ice cream should be revoked. It bears little resemblance in goodness and composition to the real thing and several of its ingredients are known to cause cancer and birth defects in test animals.*

When one considers the thousands of chemicals permitted in the production of artificial flavours and colours as well as the countless other chemicals permitted in the manufacturing process, it is easy to understand how ice cream can be made up of hundreds of chemicals. This is not readily apparent in the following excerpts. If you turn to Tables III, IV, X and XII (see *Tables of Additives According to Function) which are the lists of chemicals allowed as colouring agents, stabilizing agents, pH-adjusting agents and sequestering agents, you will see the array of substances which processors are permitted to choose from when making ice cream.*

Canadians consume more than sixty million gallons of ice cream a year. As more information becomes available concerning its harmful properties, enlightened people are enjoying the pleasure of making their own ice cream or seeking entrepreneurs who are producing real, genuine ice cream. Nothing compares to the taste of the real thing.

B.08.061. [S].
Ice Cream Mix

(a) shall be the unfrozen, pasteurized combination of cream, milk or other milk products, sweetened with **sugar, liquid sugar, invert sugar,** honey, **dextrose, glucose, corn syrup solids** of any combination of such sweeteners;
(b) may contain
  (i) egg,
  (ii) **a flavouring preparation,**
  (iii) cocoa or chocolate syrup,
  (iv) **a food colour,**
  (v) **pH adjusting agents,**

(vi) **microcrystalline cellulose or a stabilizing agent or both** in an amount that will not exceed 0.5 per cent of the ice cream made from the mix,
(vii) **a sequestering agent,**
(viii) **salt;** and
(ix) not more than 1 per cent added edible casein or edible caseinates; and

(c) shall contain not less than
(i) 36 per cent solids, and
(ii) 10 per cent milk fat or, where cocoa or chocolate syrup has been added, 8 per cent milk fat.

B.08.062. [S].
Ice Cream

(a) shall be the frozen food obtained by freezing **an ice cream mix,** with or without the incorporation of air;
(b) may contain cocoa or chocolate syrup, fruit, nuts or confections;
(c) shall contain not less than
(i) 36 per cent solids,
(ii) 10 per cent milk fat or, where cocoa or chocolate syrup, fruit, nuts, or confections have been added, 8 per cent milk fat, and
(iii) 180 grams of solids per litre of which amount not less than 50 grams shall be milk fat, or, where cocoa or chocolate syrup, fruit, nuts or confections have been added, 180 grams of solids per litre of which amount not less than 40 grams shall be milk fat; and
(d) shall contain not more than
(i) 100,000 bacteria per gram, and
(ii) 10 coliform organisms per gram, as determined by official method MFO-2.

B.08.063. [S].
Sherbet

(a) shall be the frozen food, other than ice cream or ice milk, made from a milk product;
(b) **may contain**
(i) water,
(ii) **a sweetening agent,**

(iii) fruit or fruit juice,
(iv) citric or tartaric acid,
(v) **a flavouring preparation,**
(vi) **a food colour,**
(vii) not more than 0.75 per cent **stabilizing agent,**
(viii) **a sequestering agent,**
(ix) lactose,
(x) not more than 0.5 per cent microcrystalline cellulose; and
(xi) not more than 1 per cent added edible casein or edible caseinates; and

(c) shall contain
(i) not more than 5 per cent milk solids, including milk fat, and
(ii) not less than 0.35 per cent acid determined by titration and expressed as lactic acid.

B.08.071. [S].
Ice Milk Mix

(a) shall be the unfrozen, pasteurized combination of cream, milk or other milk products, sweetened with **sugar, liquid sugar, invert sugar,** honey, **dextrose, glucose, corn syrup, corn syrup solids or any combination of such sweeteners;**

(b) may contain
(i) egg,
(ii) **a flavouring preparation,**
(iii) cocoa or chocolate syrup,
(iv) **a food colour,**
(v) **a pH-adjusting agent,**
(vi) **a stabilizing agent,** in an amount that will not result in more than 0.5 per cent stabilizing agent in the ice milk,
(vii) a sequestering agent,
(viii) added lactose, and
(ix) not more than 1.5 per cent **microcrystalline cellulose;**
(x) salt; and
(xi) not more than 1 per cent added edible casein or edible caseinates; and

(c) shall contain
(i) not less than 33 per cent solids, and
(ii) not less than 3 per cent and not more than 5 per cent milk fat.

B.08.075. [S].
Cream

(a) shall be the fatty liquid prepared from milk by separating the milk constituents in such a manner as to increase the milk fat content, and
(b) **may contain**
  (i) **a pH-adjusting agent,**
  (ii) **a stabilizing agent,** and
  (iii) in the case of **cream for whipping** that has been heat-treated above 100°C, the following ingredients and food additives:
    (A) skim milk powder in an amount not exceeding 0.25 per cent,
    (B) glucose solids in an amount not exceeding 0.1 per cent,
    (C) calcium sulphate in an amount not exceeding 0.005 per cent,
    (D) **xanthan gum** in an amount not exceeding 0.02 per cent, and
    (E) **microcrystalline cellulose** in an amount not exceeding 0.2 per cent.

B.08.077. [S].
Sour Cream

(a) shall be the product prepared by the souring of pasteurized cream with acid-producing bacterial culture and shall contain not less than 14 per cent milk fat; and
(b) may contain
  (i) milk solids,
  (ii) whey solids,
  (iii) buttermilk,
  (iv) starch in an amount not exceeding 1 per cent
  (v) salt,
  (vi) **rennet as a food enzyme** in an amount consistent with good manufacturing practice,
  (vii) **the following emulsifying, gelling, stabilizing and thickening agents:**
    (A) **algin, carob bean gum (locust bean gum), carrageenan, gelatin, guar gum, pectin or**

propylene glycol alginate or any combination thereof in an amount not exceeding 0.5 per cent,
(B) **monoglycerides, mono- and diglycerides, or any combination thereof,** in an amount not exceeding 0.3 per cent, and
(C) **sodium phosphate dibasic** in an amount not exceeding 0.05 per cent,

(viii) sodium citrate as a flavour precursor in an amount not exceeding 0.1 per cent,
(ix) milk coagulating enzyme derived from *Mucor miehei* (Cooney and Emmerson) or *Mucor pusillus Lindt*, in an amount consistent with good manufacturing practice,
(x) **Chymosin A** derived from *Escherichia coli* K-12, **GE81 (pPFZ87A),** in an amount consistent with good manufacturing practice, and
(xi) **Chymosin B** derived from *Aspergillus niger* var. *awamori*, **GCC0349 (pGAMpR)** or from *Kluyveromyces marxianus* var. *lactis*, **DS1182 (pKS105)**, in an amount consistent with good manufacturing practice.

DIVISION 9

Fats and Oils

B.09.001. [S].
Vegetable fats and oils shall be fats and oils obtained entirely from the botanical source after which they are named, shall be dry and sweet in flavour and odour and, with the exception of olive oil, **may contain emulsifying agents, Class IV preservatives, an antifoaming agent,** and ß-carotene in a quantity sufficient to replace that lost during processing, if such an addition is declared on the label.

B.09.002. [S].
Animal fats and oils shall be fats and oils obtained entirely from animals healthy at the time of slaughter, shall be dry and sweet in flavour and odour and **may contain**

(a) with the exception of milk fat and suet, **Class IV preservatives;** and
(b) with the exception of lard, milk fat and suet, **an antifoaming agent.**

B.09.004. [S].
Cotton Seed Oil

(a) shall be the oil of the seeds of cultivated *Gossypium spp.*;
(b) shall have
  (i) a relative density (20°C/water at 20°C) of not less than 0.918 and not more than 0.926,
  (ii) a refractive index (n_D 40°C) of not less than 1.458 and not more than 1.466,
  (iii) a saponification value (milligrams potassium hydroxide per gram of oil) of not less than 189 and not more than 198,
  (iv) an iodine value (Wijs) of not less than 99 and not more than 119,
  (v) an unsaponifiable matter content of not more than 15 grams per kilogram,
  (vi) a positive Halphen test,
  (vii) an acid value of not more than 0.6 milligram potassium hydroxide per gram of oil, and
  (viii) peroxide value of not more than 10 milliequivalents peroxide oxygen per kilogram of oil; and
(c) may contain oxystearin.

B.09.006. [S].
Corn Oil or Maize Oil

(a) shall be the oil of the germ or embryo of *Zea mays L.*; and
(b) shall have
  (i) a relative density (20°C/water at 20°C) of not less than 0.917 and not more than 0.925,
  (ii) a refractive index (n_D 40°C) of not less than 1.465 and not more than 1.468,
  (iii) a saponification value (milligram potassium hydroxide per gram of oil) of not less than 187 and not more than 195,
  (iv) an iodine value (Wijs) of not less than 103 and not more than 128,
  (v) an unsaponifiable matter content of not more than 28 grams per kilogram,
  (vi) an acid value of not more than 0.6 milligram potassium hydroxide per gram of oil, and
  (vii) a peroxide value of not more than 10 milliequivalents peroxide oxygen per kilogram of oil.

B.09.011. [S].

Shortening, other than butter or lard, shall be the semi-solid food prepared from fats, oils or a combination of fats and oils, may be processed by hydrogenation and may contain

(a) **Class IV preservatives;**
(b) **an antifoaming agent;**
(c) **stearyl monoglyceridyl citrate;**
(d) **monoglycerides or a combination of monoglycerides and diglycerides** of fat forming fatty acids, the weight of the monoglycerides being not more than 10 per cent and the total weight of monoglycerides and diglycerides being not more than 20 per cent of the weight of the shortening;
(e) **lactylated monoglycerides,** or a combination of lactylated monoglycerides and diglycerides of fat forming fatty acids, the total weight being not more than 8 per cent of the weight of the shortening; and
(f) **sorbitan tristearate** except that the total weight of the ingredients permitted under paragraphs (d) and (e) shall not be greater than 20 per cent of the weight of the shortening.

Authors' Note: *Regulation B.09.012 (below) was repealed in its entirety on March 19, 1997. Although the standard for the preparation of monoglycerides has been repealed, monoglycerides are still used in numerous foods. What explanation could there be for removing the standard for the preparation of a chemical that is still permitted for use in so many foods?*

B.09.012. [S].

Monoglycerides, Monoglycerides and Diglycerides shall be monoglycerides or monoglycerides and diglycerides of fat-forming fatty acids and may contain

(a) not more than 0.02 per cent glycine;
(b) acid;
(c) not more than 2.5 per cent glycerol;
(d) an antifoaming agent; and
(e) Class IV preservatives.

Authors' Note: *The following excerpt clearly illustrates that margarine is unnatural and devoid of nutrition. Compare the ingredients of margarine to those of butter and then ask yourself "Is it better to eat a plastic food or the real thing?" We'll take our chances with butter!*

B.09.016. [S].
Margarine

(a) **shall be a plastic or fluid emulsion** of water in fats, oil or fats and oil that are not derived from milk **and may have been subjected to hydrogenation;**

(b) shall contain
  - (i) not less than 80 per cent fat, oil or fat and oil calculated as fat, and
  - (ii) notwithstanding section D.01.009, not less than
    - (A) 3,300 International Units of vitamin A, and
    - (B) 530 International Units of vitamin D per 100 grams; and

(c) **may contain**
  - (i) skim milk powder, buttermilk powder or liquid buttermilk,
  - (ii) whey solids or modified whey solids,
  - (iii) protein,
  - (iv) water,
  - (v) vitamin E, if added in such an amount as will result in the finished product containing not less than 0.6 International Units of alphatocopherol per gram of linoleic acid present in the margarine,
  - (vi) **flavouring agent,**
  - (vii) **a sweetening agent,**
  - (viii) **potassium chloride and sodium chloride,**
  - (ix) **the following colouring agents:** annatto, ß-apo-8'-carotenal, canthaxanthin, carotene, ethyl ß-apo-8'-carotenoate and turmeric, as set out in Table III to section B.16.100,
  - (x) **the following emulsifying agents: lecithin, mono- and diglycerides, monoglycerides and sorbitan tristearate, as set out in Table IV to section B.16.100,**
  - (xi) **the following pH-adjusting agents: citric acid, lactic acid, potassium bicarbonate, sodium bicarbonate, potassium carbonate, sodium carbonate, sodium citrate, sodium lactate, potassium citrate, potassium hydroxide, sodium hydroxide, potassium lactate, sodium potassium tartrate and tartaric acid, as set out in Table X to section B.16.100,**

(xii) **the following preservatives: ascorbyl palmitate, ascorbyl stearate, benzoic acid, butylated hydroxyanisole, butylated hydroxytoluene, calcium sorbate, monoglyceride citrate, monoisopropyl citrate, potassium benzoate, potassium sorbate, propyl gallate, sodium benzoate, sodium sorbate and sorbic acid, as set out in Table XI to section B.16.100, and**

(xiii) **the following sequestering agents: calcium disodium ethylenediaminetetraacetate and stearyl citrate, as set out in Table XII to section B.16.100.**

DIVISION 10

Flavouring Preparations

B.10.003. [S].
(naming the flavour) Extract or **(naming the flavour) Essence** shall be a solution in ethyl alcohol, glycerol, propylene glycol or any combination of these, of sapid or odorous principles, or both, derived from the plant after which the flavouring extract or essence is named, and may contain water, a sweetening agent, food colour and a Class II preservative or Class IV preservative.

B.10.004. [S].
Artificial (naming the flavour) Extract, Artificial (naming the flavour) Essence, Imitation (naming the flavour) Extract or **Imitation (naming the flavour) Essence** shall be a flavouring extract or essence except that the flavouring principles shall be derived in whole, or in part, from sources other than the aromatic plant after which it is named.

B.10.005. [S].
(naming the flavour) Flavour

(a) shall be a preparation, other than a flavouring preparation described in section B.10.003, of sapid or odorous principles, or both, derived from the aromatic plant after which the flavour is named;

(b) **may contain** a sweetening agent, **food colour, Class II**

preservative, thaumatin, Class IV preservative or emulsifying agent; and

(c) **may have added to it** the following liquids only:
  (i) water;
  (ii) **any of, or any combination of, the following: benzyl alcohol; 1,3-butylene glycol, ethyl acetate, ethyl alcohol, glycerol, glyceryl diacetate, glyceryl triacetate, glyceryl tributyrate, isopropyl alcohol, monoglycerides and diglycerides; 1,2-propylene glycol or triethylcitrate;**
  (iii) edible vegetable oil; and
  (iv) **brominated vegetable oil, sucrose acetate isobutyrate or mixtures thereof,** when such flavour is used in citrus-flavoured or spruce-flavoured beverages.

B.10.016. [S].
Lemon Essence, Lemon Extract or **Lemon Flavour** shall be the essence, extract or flavour prepared from natural or terpeneless oil of lemon or from lemon peel and shall contain not less than 0.2 per cent citral derived from oil of lemon.

B.10.018. [S].
Orange Essence, Orange Extract or **Orange Flavour** shall be the essence, extract or flavour prepared from sweet orange peel, oil of sweet orange or terpeneless oil of sweet orange, and shall correspond in flavouring strength to an alcoholic solution containing 5 per cent by volume of oil of sweet orange, the volatile oil obtained from the fresh peel of *Citrus aurantium L.*, that shall have an optical rotation, at a temperature of 25°C, of not less than +95 (using a tube 100 millimetres in length).

B.10.019. [S].
Peppermint Essence, Peppermint Extract or **Peppermint Flavour** shall be the essence, extract or flavour prepared from peppermint or oil of peppermint, obtained from the leaves and flowering tops of *Mentha piperita L.*, or of *Mentha arvensis De.C.*, var. *piperascens*, Holmes, and shall correspond in flavouring strength to an alcoholic solution of not less than 3 per cent by volume of oil of peppermint, containing not less than 50 per cent free and combined menthol.

B.10.026. [S].
Vanilla Extract, Vanilla Essence or **Vanilla Flavour**

(a) shall be the essence, extract, or flavour prepared from the vanilla bean, the dried, cured fruit of *Vanilla planifolia*, Andrews, or *Vanilla tahitensia*, J. W. Moore,

(b) shall contain in 100 mL, regardless of the method of extraction, at least the quantity of soluble substances in their natural proportions that are extractable, according to official method FO-17,

  (i) not less than 10 g of vanilla beans, where the beans contain 25 per cent or less moisture, and

  (ii) not less than 7.5 g of vanilla beans on the moisture-free basis, where the beans contain more than 25 per cent moisture; and

(c) notwithstanding sections B.10.003 and B.10.005, shall not contain added colour.

DIVISION 11

Fruits, Vegetables, Their Products and Substitutes

B.11.001.
In this division

(a) an acid ingredient means

  (i) citric, malic or tartaric acid,

  (ii) lemon or lime juice, or

  (iii) vinegar;

(b) a sweetening ingredient means **sugar, invert sugar,** honey, **dextrose, glucose or glucose solids or any combination thereof** in dry or liquid form; and

(c) fruit juice means the unfermented liquid expressed from sound ripe fresh fruit, and includes any such liquid that is heat treated and chilled.

B.11.001.1.
No person shall sell any fresh fruit or vegetable that is intended to be consumed raw, **except grapes, if sulphurous acid** or any salt thereof has been added thereto.

Vegetables

B.11.002. [S].
Canned (naming the vegetable)

(a) shall be the product obtained by heat processing the named fresh vegetable after it has been properly prepared;
(b) shall be packed in hermetically sealed containers;
(c) may contain
  (i) **a sweetening ingredient,**
  (ii) salt,
  (iii) water, and
  (iv) **a firming agent;** and
(d) **may contain**
  (i) in the case of canned green beans and canned wax beans, pieces of green peppers, red peppers and tomato in an amount not exceeding 15 per cent of the final product, and dill seasonings and vinegar,
  (ii) **in the case of canned peas,** garnishes composed of one or more of lettuce, onions, carrots, and pieces of green or red peppers in an amount not exceeding 15 per cent of the total drained vegetable ingredient, aromatic herbs, spices and seasonings, stock or juice of vegetables and aromatic herbs, **calcium hydroxide** in an amount not exceeding 0.01 per cent of the final product **and magnesium hydroxide** in an amount not exceeding 0.05 per cent of the final product,
  (iii) in the case of
    (A) canned asparagus, acetic acid, citric acid, malic acid and tartaric acid at levels consistent with good manufacturing practice, and
    (B) canned white asparagus, acetic acid, citric acid, malic acid, tartaric acid and ascorbic acid at levels consistent with good manufacturing practice,
  (iv) in the case of **asparagus packed in glass containers** or fully lined (lacquered) **cans, stannous chloride** in an amount not exceeding 25 parts per million, **calculated as tin,**
  (v) in the case of canned artichokes, canned bean sprouts and canned onions, citric acid at levels consistent with good manufacturing practice to be used as a pH adjusting agent,

(vi) in the case of **canned ripe lima beans** (butter beans) and canned pinto beans, **calcium disodium ethylenediaminetetraacetate** in an amount not exceeding 130 parts per million,

(vi.1) in the case of **canned fava beans, calcium disodium ethylenediaminetetraacetate** in an amount not exceeding 365 parts per million,

(vii) in the case of **canned red kidney beans, canned chick peas (garbanzo beans) and canned black-eye peas, disodium ethylenediaminetetraacetate** in an amount not exceeding 150 parts per million, and

(viii) in the case of canned asparagus, canned green beans, canned wax beans and canned peas

(A) butter or other edible animal or vegetable fats or oils, but if butter is added it shall be not less than 3 per cent of the final product,

(B) natural or enzymatically or physically modified starches when used with butter or other edible animal or vegetable fats and oils,

(C) **acacia gum, algin, carrageenan, furcelleran, guar gum and propylene glycol alginate** used with butter or other edible animal or vegetable fats or oils in an amount not exceeding 1 per cent, **singly or in any combination,** of the final product, and

(D) characterizing sauces, seasonings or **flavouring agents** if it is included in the common name of the product.

B.11.005. [S].
Tomatoes or Canned Tomatoes

(a) shall be the product made by heat processing properly prepared fresh ripe tomatoes;

(b) **may contain**

(i) **a sweetening ingredient** in dry form,

(ii) salt,

(iii) **a firming agent,**

(iv) citric acid, and

(v) spice or other seasoning; and

(c) **shall contain not less than 50 per cent** drained **tomato solids,** as determined by official method FO-18.

B.11.007. [S].
Tomato Juice shall be the unconcentrated, pasteurized liquid containing a substantial portion of fine tomato pulp extracted from sound, ripe, whole tomatoes from which all stems and objectionable portions have been removed by any method that does not add water to the liquid **and may contain** salt and **a sweetening ingredient in dry form.**

B.11.009. [S].
Tomato Paste shall be the product made by evaporating a portion of the water from tomatoes or sound tomato trimmings, **may contain salt and Class II preservatives and shall contain not less than 20 per cent tomato solids,** as determined by official method FO-19.

B.11.010. [S].
Concentrated Tomato Paste shall be tomato paste containing **not less than 30 per cent tomato solids,** as determined by official method FO-19.

B.11.011. [S].
Tomato Pulp shall be the heat processed product made from whole, ripe tomatoes or sound tomato trimmings concentrated to yield a product with a specific gravity of not less than 1.050 and **may contain salt and a Class II preservative.**

B.11.012. [S].
Tomato Purée shall be the heat processed product made from whole, ripe tomatoes, with the skins and seeds removed, concentrated to yield a product with a specific gravity of not less than 1.050 and **may contain salt and a Class II preservative.**

B.11.014. [S].
Tomato Catsup, Catsup or products whose common names are variants of the word Catsup

(a) shall be the heat processed product made from the juice of red-ripe tomatoes or sound tomato trimmings from which skins and seeds have been removed;

(b) **shall contain**

(i) vinegar,

(ii) salt,

(iii) **seasoning,** and

(iv) **a sweetening ingredient; and**

(c) **may contain**
  (i) **a Class II preservative, and**
  (ii) **food colour.**

B.11.050. [S].
Olives shall be the plain or stuffed fruit of the olive tree, and may contain
(a) vinegar,
(b) salt,
(c) **a sweetening ingredient,**
(d) **spices,**
(e) **seasonings,**
(f) lactic acid; and
(g) in the case of ripe olives, **ferrous gluconate,**
(h) sorbic acid and its potassium or sodium salt, and
(i) **calcium chloride.**

B.11.051. [S].
Pickles and Relishes shall be the product prepared from vegetables or fruits with salt and vinegar, and **may contain**
(a) **spices,**
(b) **seasonings,**
(c) **sugar, invert sugar, dextrose or glucose,** in dry or liquid form,
(d) **food colour,**
(e) **a Class II preservative,**
(f) **a firming agent,**
(g) **polyoxyethylene (20) sorbitan monooleate** in an amount not exceeding 0.05 per cent,
(h) lactic acid,
(i) vegetable oils, and
(j) **in the case of relishes and mustard pickles, a thickening agent.**

Fruits

B.11.101. [S].
Canned (naming the fruit)
(a) shall be the product prepared by heat processing the named fresh fruit after it has been properly prepared;

(b) shall be packed in hermetically sealed containers; and
(c) **may contain**
 (i) **a sweetening ingredient,**
 (ii) water,
 (iii) fruit juice, fruit juice from concentrate, concentrated fruit juice or any combination thereof,
 (iv) **in the case of canned pears,** citric acid, malic acid, L-tartaric or lactic acid at a level sufficient to maintain pH 4.2 to 4.5, lemon juice, spices, spice oils, mint and a **flavouring preparation** other than that which simulates the flavour of canned pears,
 (v) **in the case of canned apples, a firming agent,**
 (vi) **in the case of canned applesauce,** citric acid and malic acid at a level sufficient to maintain pH 4.2 to 4.5, ascorbic acid and isoascorbic acid provided the total does not exceed 150 parts per million, **spices, salt and a flavouring preparation** other than that which simulates the flavour of canned applesauce,
 (vii) **in the case of canned grapefruit,** citric acid at a level sufficient to maintain pH 4.2 to 4.5, lemon juice, **calcium chloride and calcium lactate** provided the total calcium content, whether naturally present or added, does not exceed 0.035 per cent, **spices and a flavouring preparation** other than that which simulates the flavour of canned grapefruit,
 (viii) in the case of canned mandarin oranges, citric acid at a level sufficient to maintain pH 4.2 to 4.5,
 (ix) **in the case of canned peaches,** L-ascorbic acid at a level not to exceed 550 parts per million, spices, peach pits and peach kernels intended for flavour development and **a flavouring preparation** other than that which simulates the flavour of canned peaches,
 (x) **in the case of canned pineapple,** citric acid at a level sufficient to maintain pH 4.2 to 4.5, **spices, spice oils,** mint, **dimethylpolysiloxane** not to exceed 10 parts per million when pineapple juice is used as a packing medium **and a flavouring preparation** other than that which simulates the flavour of canned pineapple,
 (xi) **in the case of canned plums, a flavouring preparation** other than that which simulates the flavour of canned plums, and

(xii) in the case of canned strawberries, citric acid, lactic acid, malic acid or L-tartaric acid at a level sufficient to maintain pH 4.2 to 4.5.

Fruit Juices

B.11.120. [S].
(naming the fruit) Juice

(a) shall be the juice obtained from the named fruit; and
(b) may contain **a sweetening ingredient** in dry form, **a Class II preservative, amylase, cellulase and pectinase.**

B.11.121.
Notwithstanding section B.11.120, the fruit juice prepared from any fruit named in any of sections B.11.123 to B.11.128A shall conform to the standard prescribed for that fruit juice in that section.

B.11.123. [S].
Apple Juice

(a) shall be the fruit juice obtained from apples;
(b) may contain **a Class II preservative,** vitamin C, **amylase, cellulase and pectinase;**
(c) shall have a specific gravity of not less than 1.041 and not more than 1.065; and
(d) shall contain in 100 millilitres measured at a temperature of 20°C, not less than 0.24 gram and not more than 0.60 gram of ash of which not less than 50 per cent shall be **potassium carbonate.**

B.11.124. [S].
Grape Juice

(a) shall be the fruit juice obtained from grapes;
(b) shall have a specific gravity of not less than 1.040 and not more than 1.124;
(c) shall contain, before the addition of **a sweetening ingredient,** in 100 millilitres measured at a temperature of 20°C,
   (i) not less than 0.20 gram and not more than 0.55 gram of ash, and
   (ii) not less than 0.015 gram and not more than 0.070 gram of **phosphoric acid calculated as phosphorous pentoxide;** and

(d) **may contain a pH-adjusting agent, a sweetening ingredient** in dry form, **a Class II preservative,** vitamin C, **amylase, cellulase and pectinase.**

B.11.128. [S].
Orange Juice

(a) shall be fruit juice obtained from clean, sound, mature oranges;
(b) shall
  (i) contain not less than 1.20 milliequivalents of free amino acids per 100 millilitres,
  (ii) contain not less than 115 milligrams of potassium per 100 millilitres, as determined by official method FO-22, and
  (iii) have an absorbance value for total polyphenolics of not less than 0.380, as determined by official method FO-23,
(c) shall, before the addition of sugar, invert sugar, dextrose or glucose solids,
  (i) have a Brix reading of not less than 9.7°, as determined by official method FO-24, and
  (ii) contain not less than 0.5 per cent and not more than 1.8 per cent of acid by weight calculated as anhydrous citric acid, as determined by official method FO-25;
(d) **may contain orange essences, orange oils** and orange pulp adjusted in accordance with good manufacturing practice; and
(e) **may contain sugar, invert sugar, dextrose** in dry form, glucose solids, **a Class II preservative, amylase, cellulase and pectinase.**

Authors' Note: *Would it not be much more nutritious for your children to squeeze an orange and let them drink pure orange juice?*

B.11.128A. [S].
Pineapple Juice

(a) shall be the fruit juice obtained from pineapple; and
(b) **may contain a sweetening ingredient** in dry form, **a Class II preservative,** vitamin C, **amylase, cellulase, pectinase and an antifoaming agent.**

B.11.130. [S].
(1) Concentrated (naming the fruit) Juice

(a) shall be fruit juice that is concentrated to at least one half of its original volume by the removal of water;

(b) may contain
  (i) vitamin C,
  (ii) **food colour,**
  (iii) **stannous chloride,**
  (iv) **a sweetening ingredient,** and
  (v) **a Class II preservative;** and
(c) may have added to it, for the purpose of adjustment in accordance with good manufacturing practice, all or any of the following, namely,
  (i) essence, oil and pulp from the named fruit, and
  (ii) water.

(2) Sub-paragraphs (1)(b)(i), (ii), (iii) and (v) do not apply in respect of frozen concentrated orange juice.

Jams

B.11.201. [S].
(naming the fruit) Jam

(a) shall be the product obtained by processing fruit, fruit pulp, or canned fruit, by boiling to a suitable consistency with water and a sweetening ingredient
(b) shall contain not less than
  (i) 45 per cent of the named fruit, and
  (ii) 66 per cent water soluble solids estimated by the refractometer;
(c) **may contain**
  (i) such amount of added pectin, pectinous preparation, or acid ingredient as reasonably compensates for any deficiency in the natural pectin content or acidity of the named fruit,
  (ii) **a Class II preservative,**
  (iii) **a pH-adjusting agent,** and
  (iv) **an antifoaming agent;** and
(d) shall not contain apple or rhubarb.

B.11.202. [S].
(naming the fruit) Jam with Pectin

(a) shall be the product obtained by processing fruit, fruit pulp, or canned fruit by boiling to a suitable consistency with water and a sweetening ingredient;

(b) shall contain
  (i) not less than 27 per cent of the named fruit,
  (ii) not less than 66 per cent water soluble solids as estimated by the refractometer, and
  (iii) pectin or pectinous preparations;
(c) **may contain**
  (i) such amount of acid ingredients as reasonably compensates for any deficiency in the natural acidity of the named fruit,
  (ii) **food colour,**
  (iii) **a Class II preservative,**
  (iv) **a pH-adjusting agent, and**
  (v) **an antifoaming agent,** and
(d) shall not contain apple or rhubarb.

B.11.221. [S].
(naming the citrus fruit) Marmalade with Pectin
(a) shall be the food of jelly-like consistency made from any combination of peel, pulp or juice of the named citrus fruit by boiling with water and a sweetening ingredient;
(b) shall contain
  (i) not less than 27 per cent of any combination of peel, pulp or juice of the named citrus fruit,
  (ii) not less than 65 per cent water soluble solids as estimated by the refractometer, and
  (iii) pectin or pectinous preparation; and
(c) may contain
  (i) such amount of acid ingredient as reasonably compensates for any deficiency in the natural acidity of the citrus fruit used in its preparation,
  (ii) **a Class II preservative,**
  (iii) **a pH-adjusting agent, and**
  (iv) **an antifoaming agent.**

B.11.241. [S].
(Naming the fruit) Jelly with Pectin
(a) shall be the gelatinous food, free of seeds and pulp, made from the named fruit, the juice of the named fruit or a concentrate of the juice of the named fruit which has been boiled with water and a sweetening ingredient;

(b) shall contain
  (i) not less than the equivalent of 32 per cent juice of the named fruit,
  (ii) not less than 62 per cent water soluble solids, as estimated by the refractometer, and
  (iii) pectin or pectinous preparation; and
(c) may contain
  (i) such amount of acid ingredient as reasonably compensates for any deficiency in the natural acidity of the named fruit,
  (ii) juice of another fruit,
  (iii) **a gelling agent,**
  (iv) **food colour,**
  (v) **a Class II preservative,**
  (vi) **a pH-adjusting agent, and**
  (vii) **an antifoaming agent.**

B.11.260. [S].
Boiled Cider shall be the liquid expressed from whole apples, apple cores, apple trimmings or apple culls and concentrated by boiling.

DIVISION 12

Prepackaged Water and Ice

B.12.001. [S].
Water represented as mineral water or spring water,
(a) shall be potable water obtained from an underground source but not obtained from a public community water supply;
(b) shall not contain any coliform bacteria, as determined by official method MFO-9,
(c) shall not have its composition modified through the use of any chemicals; and
(d) notwithstanding paragraph (c), may contain
  (i) added carbon dioxide,
  (ii) added fluoride, if the total fluoride ion content thereof does not exceed 1 part per million, and
  (iii) added ozone.

B.12.002.

The principal display panel of the label on a container of water represented as mineral water or spring water shall carry a statement

(a) of the geographical location of the underground source from which it is obtained;

(b) of the total dissolved mineral salt content expressed in parts per million;

(c) of the total fluoride ion content expressed in parts per million; and

(d) of any addition of fluoride or ozone thereto.

DIVISION 13

Grain and Bakery Products

Authors' Note: *The following excerpt shows how white flour has become one of the most adulterated ingredients in existence today. White flour is bleached to eliminate its natural creamy colour. Unfortunately, this process is very destructive to the nutrients in the flour. Present-day milling eliminates at least twenty nutrients while only four or five are put back into "Enriched Flour" and listed on the label. This leads the consumer to believe the product is nutritious. Nothing can replace the original nutritional quality of the flour. We object to the use of the term "enriched" by the very people who have robbed it of those qualities in the first place. As if this were not enough, we now find that Health Canada permits flour to be irradiated with Cobalt-60* (see *Irradiation Table in Division 26). Check with your local health food store to see if they carry unbleached and non-irradiated flour.*

B.13.001. [S].

Flour, White Flour, Enriched Flour or **Enriched White Flour**

(a) shall be the food prepared by the grinding and bolting through cloth having openings not larger than those of woven wire cloth designated "149 microns (No. 100)," of cleaned milling grades of wheat;

(b) shall be free from bran coat and germ to such an extent that the percentage of ash therein, before the addition of any other material permitted by this section calculated on a moisture free basis, does not exceed 1.20 per cent;

(c) shall have a moisture content of not more than 15 per cent;
(d) shall contain in 100 grams of flour
 (i) not less than 0.44 and not more than 0.77 milligram of thiamine,
 (ii) not less than 0.27 and not more than 0.48 milligram of riboflavin,
 (iii) not less than 3.5 and not more than 6.4 milligrams of niacin or niacinamide, and
 (iv) not less than 2.9 and not more than 4.3 milligrams of iron,
(e) **may contain**
 (i) malted wheat flour,
 (ii) malted barley flour in an amount not exceeding 0.50 per cent of the weight of the flour,
 (iii) **amylase, amylase (maltogenic), bromelain, glucoamylase, lactase, lipoxidase, pentosanase, protease and pullulanase,**
 (iv) **chlorine,**
 (v) **chlorine dioxide,**
 (vi) **benzoyl peroxide** in an amount not exceeding 150 parts by weight for each 1 million parts of flour, with or without not more than 900 parts by weight for each 1 million parts of flour of one **or a mixture of two or more of calcium carbonate, calcium sulphate, dicalcium phosphate, magnesium carbonate, potassium aluminum sulphate, sodium aluminum sulphate, starch and tricalcium phosphate** as carriers of the benzoyl peroxide,
 (vii) Revoked by P.C. 1994-362 of March 1, 1994,
 (viii) **ammonium persulphate** in an amount not exceeding 250 parts by weight for each 1 million parts of flour,
 (ix) **ammonium chloride** in an amount not exceeding 2,000 parts by weight for each 1 million parts of flour,
 (x) **acetone peroxide,**
 (xi) **azodicarbonamide** in an amount not exceeding 45 parts by weight for each 1 million parts of flour,
 (xii) ascorbic acid in an amount not exceeding 200 parts by weight for each 1 million parts of flour,
 (xiii) l-cysteine (hydrochloride) in an amount not exceeding 90 parts by weight for each 1 million parts of flour,

(xiv) **monocalcium phosphate** in an amount not exceeding 7,500 parts by weight for each 1 million parts of flour, and

(xv) in 100 grams of flour

(A) not less than 0.25 and not more than 0.31 milligram of vitamin B_6,

(B) not less than 0.04 and not more than 0.15 milligram of folic acid,

(C) not less than 1.0 and not more than 1.3 milligrams of *d*-pantothenic acid, and

(D) not less than 150 and not more than 190 milligrams of magnesium;

(f) **may contain calcium carbonate,** edible bone meal, **chalk (B.P.), ground limestone or calcium sulphate** in an amount that will provide in 100 grams of flour not less than 110 and not more than 140 milligrams of calcium.

Authors' Note: *Although the refining of whole wheat flour results in a substantial loss of nutrients, it remains superior to white flour. As the following regulation indicates, several additives are permitted in the preparation of wholewheat flour. It is unfortunate that such a potentially excellent and widely used ingredient has been permitted to be so adulterated. Quality whole wheat flour, available in health food stores, is nutritionally superior to commercial brands due to the care these manufacturers take to retain a portion of the wholesome natural nutrients of the wheat. Having said all of this, we must point out that it is permitted to be irradiated with Cobalt-60* (see *Irradiation Table in Division 26).*

B.13.005. [S].

Whole Wheat Flour or Entire Wheat Flour

(a) shall be the food prepared by the grinding and bolting of cleaned, milling grades of wheat from which a part of the outer bran or epidermis layer may have been separated;

(b) shall contain the natural constituents of the wheat berry to the extent of not less than 95 per cent of the total weight of the wheat from which it is milled;

(c) shall have

(i) an ash content, calculated on a moisture-free basis, of not less than 1.25 per cent and not more than 2.25 per cent,

(ii) a moisture content of not more than 15 per cent, and
(iii) such a degree of fineness that not less than 90 per cent bolts freely through a No. 8 (2380 micron) sieve, and not less than 50 per cent through a No. 20 (840 micron) sieve,

(d) **may contain**

(i) malted wheat flour,
(ii) malted barley flour in an amount not exceeding 0.50 per cent of the weight of the flour,
(iii) **amylase, amylase (maltogenic), bromelain, glucoamylase, lactase, lipoxidase, pentosanase, protease and pullulanase,**
(iv) **chlorine,**
(v) **chlorine dioxide,**
(vi) **benzoyl peroxide** in an amount not exceeding 150 parts by weight for each 1 million parts of flour, with or without not more than 900 parts by weight for each 1 million parts of flour of one **or a mixture of two or more of calcium carbonate, calcium sulphate, dicalcium phosphate, magnesium carbonate, potassium aluminum sulphate, sodium aluminum sulphate, starch and tricalcium phosphate** as carriers of the benzoyl peroxide,
(vii) Revoked March 1, 1994.
(viii) **ammonium persulphate** in an amount not exceeding 250 parts by weight for each 1 million parts of flour,
(ix) **ammonium chloride** in an amount not exceeding 2,000 parts by weight for each 1 million parts of flour,
(x) **azodicarbonamide** in an amount not exceeding 45 parts by weight for each 1 million parts of flour,
(xi) acetone peroxide,
(xii) ascorbic acid in an amount not exceeding 200 parts by weight for each 1 million parts of flour,
(xiii) *l*-cysteine (hydrochloride) in an amount not exceeding 90 parts by weight for each 1 million parts of flour.

B.13.010. [S].
Rice shall be the hulled or hulled and polished seed of the rice plant and, in the case of hulled and polished seeds, may be coated with magnesium silicate, talc and glucose.

Authors' Note: *For centuries, bread consisted of natural flour, liquid and yeast. The bread you now buy in your local store may contain any of the chemicals listed in this excerpt. Since Section B.01.009 exempts some of the components of bread from being listed on a label, you will never know what your bread actually contains. The over-processing of flour and the addition of such a long list of chemical additives has contributed to an alarming decrease in the nutritive quality of modern bread from that of the wholesome breads once prepared at home. In fact, white bread is one of the most adulterated and undesirable foods in existence today. An alternative would be to buy commercially prepared whole wheat bread from a specialty bakery which uses quality whole wheat flour. Even better, prepare your own whole wheat bread at home. It is easy now, especially if you use a bread machine. You simply throw all the ingredients in the machine and it kneads the dough, raises it then cooks it. When the timer goes off, you remove a loaf of the most delicious, wholesome bread you have ever tasted. You will never go back to store-bought bread.*

B.13.021. [S].

Bread or **White Bread** shall be the food made by baking a yeast-leavened dough prepared with flour and water and may contain

(a) salt;
(b) shortening, lard, butter or margarine;
(c) milk or milk product;
(d) whole egg, egg-white, egg-yolk, (fresh, dried, or frozen);
(e) **a sweetening agent;**
(f) malt syrup, malt extract or malt flour;
(g) inactive dried yeast of the genus *Saccharomyces cerevisiae* in an amount not greater than 2 parts by weight for each 100 parts of flour used;
(h) **amylase, bromelain, glucoamylase, lactase, lipoxidase, pentosanase, protease and pullulanase;**
(i) subject to section B.13.029, one or more of the following in a total amount not exceeding 5 parts by weight per 100 parts of flour used, namely, whole wheat flour, entire wheat flour, graham flour, gluten flour, wheat meal, wheat starch, non-wheat flour, non-wheat meal or non-wheat starch, any of which may be wholly or partially dextrinized;
(j) other parts of the wheat berry;
(k) lecithin or ammonium salt of phosphorylated glyceride;
(l) **monoglycerides and diglycerides** of fat-forming fatty acids;

(m) **ammonium chloride, ammonium sulphate, calcium carbonate, calcium lactate, diammonium phosphate, dicalcium phosphate, monoammonium phosphate or any combination thereof** in an amount not greater than 0.25 parts by weight of all such additives for each 100 parts of flour used;
(n) **monocalcium phosphate** in an amount not greater than 0.75 parts by weight for each 100 parts of flour used;
(o) **calcium peroxide, ammonium persulphate, potassium persulphate or any combination thereof** in an amount not greater than 0.01 part by weight of all such additives for each 100 parts of flour used;
(p) acetone peroxide;
(q) vinegar;
(r) **Class III preservative;**
(s) **food colour;**
(t) **calcium stearoyl-2-lactylate or sodium stearoyl-2-lactylate** in an amount not greater than 0.375 parts by weight for each 100 parts of flour used;
(u) *l*-cysteine (hydrochloride) in an amount not greater than 0.009 parts by weight for each 100 parts of flour used;
(v) **calcium sulphate** in an amount not greater than 0.5 parts by weight for each 100 parts of flour used;
(w) **sodium stearyl fumarate** in an amount not greater than 0.5 parts by weight for each 100 parts of flour used;
(x) ascorbic acid in an amount not greater than 0.02 parts by weight for each 100 parts of flour used;
(y) lactic acid;
(z) **azodicarbonamide** in an amount not exceeding 45 parts by weight for each 1 million parts of flour;
(aa) **calcium iodate, potassium iodate or any combination thereof** in an amount not greater than 45 parts by weight of all such additives for each 1 million parts of flour, and
(bb) **acetylated tartaric acid esters of mono- and diglycerides** in an amount not greater than 0.6 parts by weight for each 100 parts of flour used.

Authors' Note: *The key words in the following excerpt are "shall be bread"* (see *Bread B.13.021). Manufacturers can use the same additive-laden ingredients in whole wheat bread as they use in white bread with the exception* *that whole wheat flour is used in place of white flour. This is not readily apparent in the following excerpt unless you go back and see what "bread" is made with. Most consumers don't mind paying extra for whole wheat products because they consider them healthier. Read on and decide if you are getting what you are paying for! Then come up with an alternative.*

B.13.026. (S).
(naming the percentage) Whole Wheat Bread

(a) shall
- (i) be bread in the making of which the named percentage of the flour used shall be whole wheat flour, and
- (ii) contain not less than 60 per cent whole wheat flour in relation to the total flour used; and

(b) **may**
- (i) **contain caramel,** and
- (ii) where it contains not less than 6 parts by weight of milk solids per 100 parts of the total enriched flour and whole wheat flour used, be described by the common name "(naming the percentage) whole wheat milk bread."

B.13.027. (S).
Brown Bread shall be bread **coloured** by the use of whole wheat flour, graham flour, bran, molasses or **caramel.**

DIVISION 14

Meat, Its Preparation and Products

B.14.001.
In this Division
"animal" means any animal used as food, but does not include marine and fresh water animals;

"filler" means any vegetable material (except tomato or beetroot), milk, egg, yeast or any derivative or combination thereof that is acceptable as food.

B.14.002. [S].
Meat shall be the edible part of the skeletal muscle of an animal that was healthy at the time of slaughter, or muscle **that is found in the**

tongue, diaphragm, heart or esophagus, and may contain accompanying and overlying fat together with the portions of bone, skin, sinew, nerve and blood vessels that normally accompany the muscle tissue and are not separated from it in the process of dressing, but does not include muscle found in the lips, snout, scalp or ears.

B.14.003. [S].
Meat by-product shall be an edible part of an animal, other than meat, that has been derived from one or more animals that were healthy at the time of slaughter.

Authors' Note: *Health Canada previously allowed food manufacturers to add only the colours "annatto" and "caramel" to meat. As of November 20, 1997, that law has changed. It is definitely the manufacturers and not the consumers who benefit from this change. You have to read these convoluted regulations very carefully. This one, like many others, sounds as if you cannot add the substances listed, but in (c), meat processors are given permission to use any of the colours listed there. We don't know about you, but we sure don't want them adding "allura red" or "sunset yellow" to our meat. There is absolutely no benefit to the consumer from either of these colours. If meat is fresh, meat packers don't need to add colour.*

B.14.004.
Meat, meat by-products or preparations thereof are adulterated if any of the following substances or class of substances are present therein or have been added thereto;

(a) mucous membranes, any organ or portions of the genital system, black gut, spleens, udders, lungs or any other organ or portion of animal that is not commonly sold as an article of food;
(b) preservatives other than those provided for in this Division; or
(c) colour other than annatto, allura red and sunset yellow FCF, where provided for in this Division, and caramel.

B.14.005. [S].
Prepared meat or **a prepared meat by-product** shall be any meat or any meat by-product, respectively, whether comminuted or not, to which has been added any ingredient permitted by these Regulations, or which has been preserved, placed in a hermetically-sealed container or cooked, and may contain

(a) in the case of prepared hams, shoulders, butts, picnics and backs, gelatin,
(b) in the case of partially defatted pork fatty tissue and partially defatted beef fatty tissue, **a Class IV preservative,** and
(c) where a minimum total protein content or a minimum meat protein content is prescribed in this Division, phosphate salts that do not when calculated as **sodium phosphate, dibasic,** exceed the maximum level provided therefor in Table XII to section B.16.100 **and** that are **one or more of the following** phosphate salts, namely,
 (i) **sodium acid pyrophosphate,**
 (ii) **sodium hexametaphosphate,**
 (iii) **sodium phosphate, dibasic,**
 (iv) **sodium phosphate, monobasic,**
 (v) **sodium pyrophosphate, tetrabasic,**
 (vi) **sodium tripolyphosphate,**
 (vii) **potassium phosphate, monobasic,**
 (viii) **potassium phosphate, dibasic, and**
 (ix) **potassium pyrophosphate, tetrabasic.**

B.14.007. [S].
Meat Binder or **(naming the meat product) Binder** shall be a filler with any combination of salt, sweetening agents, spices or other seasonings (except tomato), egg, egg albumen, and
(a) where sold for use in preserved meat and preserved meat by-product, may contain **potassium nitrate, potassium nitrite, sodium nitrate, sodium nitrite,** ascorbic acid, sodium ascorbate, erythorbic acid, sodium erythorbate and sodium carbonate, if the nitrate or nitrite salts or both are packaged separately from any spice or seasoning;
(b) where sold for use in prepared meat or meat by-product in which a gelling agent is a permitted ingredient, may contain **a gelling agent;**
(c) where sold for use in fresh, uncooked sausage, may contain artificial maple flavour, and
(d) may contain **an anticaking agent.**

Authors' Note: *The age-old process of smoking meat to preserve it has been practically discarded by modern food technology which now employs pumping pickle or cover pickle laced with chemicals. This method allows industry to* *preserve meat at a much lower cost; however, it is uncertain what effect the combinations of these chemicals have on the human body. If this meat is then turned into such products as wieners, sausages and sliced meats, the number of chemicals in the finished product will have risen drastically. Genuine, naturally smoked meats are still available in local farmers' markets and are so superior in taste and content that their mass-produced, chemically treated counterparts don't even compare.*

B.14.009. [S].

Pumping pickle, cover pickle and dry cure employed in the curing of preserved meat or preserved meat by-product may contain

(a) **Class I preservatives** if the nitrate or nitrite salts or both are packaged separately from any spice or seasoning;
(b) citric acid, sodium citrate or vinegar;
(c) sweetening agents, including maple sugar and maple syrup;
(d) **liquid smoke flavour, liquid smoke flavour concentrate,** salt, seasonings, spices, spice extracts, spice oils or spice oleoresins;
(e) sodium bicarbonate, sodium hydroxide or potassium hydroxide;
(f) in the case of pumping pickle for cured pork, beef and lamb cuts, **disodium phosphate, monosodium phosphate, sodium hexametaphosphate, sodium tripolyphosphate, tetrasodium pyrophosphate and sodium acid pyrophosphate** in such amount calculated as disodium phosphate, as will result in the finished product containing not more than 0.5 per cent added phosphate;
(g) in the case of pumping pickle for cured beef cuts, **enzymes,** if the principal display panel of the label of the cured beef carries, immediately preceding or following the common name, the statement "Tenderized with (naming the proteolytic enzyme or enzymes)";
(h) in the case of dry cure, **an anticaking agent or a humectant;** and
(i) in the case of pumping pickle
    (i) for cured pork hams, shoulders and backs, artificial maple flavour, and
    (ii) for cured pork bellies, **artificial maple flavour** and an orange flavour that meet the standard prescribed in section B.10.005.

B.14.015. [S].
Regular Ground Beef shall be beef meat processed by grinding and shall contain **not more than 30 per cent beef fat,** as determined by official method FO-33.

B.14.015A. [S].
Medium Ground Beef shall be beef meat processed by grinding and shall contain **not more than 23 per cent beef fat** as determined by official method FO-33.

B.14.015B. [S].
Lean Ground Beef shall be beef meat processed by grinding and shall contain **not more than 17 per cent beef fat,** as determined by official method FO-33.

B.14.020. [S].
Solid cut meat shall be
- (a) a whole cut of meat; or
- (b) a product consisting of pieces of meat of which at least 80 per cent weigh at least 25 g each.

B.14.021.
(1) No person shall sell solid cut meat to which phosphate salts or water has been added unless
- (a) in the case of meat, other than side bacon, Wiltshire bacon, pork jowls, salt port and salt beef, the meat
  - (i) where cooked, contains a meat protein content of not less than 12 per cent, and
  - (ii) where uncooked, contains a meat protein content of not less than 10 per cent; and
- (b) that meat contains phosphate salts that do not, when calculated as **sodium phosphate, dibasic,** exceed the maximum level provided therefor in Table XII to section B.16.100 **and** that are **one or more of the following** phosphate salts, namely,
  - (i) **sodium acid pyrophosphate,**
  - (ii) **sodium hexametaphosphate,**
  - (iii) **sodium phosphate, dibasic,**
  - (iv) **sodium phosphate, monobasic,**
  - (v) **sodium pyrophosphate, tetrabasic,**

(vi) **sodium tripolyphosphate,**
(vii) **potassium phosphate, monobasic,**
(viii) **potassium phosphate, dibasic, and**
(ix) **potassium pyrophosphate, tetrabasic**

B.14.031. [S].
Preserved Meat or **Preserved Meat By-product** shall be cooked or uncooked meat or meat by-product that is salted, dried, pickled, corned, cured or smoked, may be glazed and may contain
(a) **Class I preservative;**
(b) **sweetening agents;**
(c) spices and seasonings, except tomato;
(d) vinegar;
(e) alcohol;
(f) smoke flavouring or **artificial smoke flavouring** if the principal display panel of the label carries, immediately preceding or following the common name, the statement "Smoke Flavouring Added" or "Artificial Smoke Flavouring Added," whichever is applicable;
(g) in the case of cured pork hams, shoulders, backs and bellies, **artificial maple flavour** if the principal display panel of the label carries, immediately preceding or following the common name, the statement "Artificial Maple Flavour Added;"
(gg) in the case of cured pork bellies, an added orange flavour that meets the standard prescribed in section B.10.005 if the principal display panel of the label carries, immediately preceding or following the common name, the statement "Orange Flavour Added;"
(h) in the case of cured pork, beef and lamb cuts prepared with the aid of pumping pickle, **disodium phosphate, monosodium phosphate, sodium hexametaphosphate sodium tripolyphosphate, tetrasodium pyrophosphate and sodium acid pyrophosphate** in such amount calculated as disodium phosphate, as will result in the finished product containing not more than 0.5 per cent added phosphate; and,
(i) in the case of tocino, annatto in such amount as will result in the finished product containing not more than 0.1 per cent annatto, if annatto is shown, by the word "annatto," in the list of ingredients on the label.

B.14.032. [S].
Sausage or **Sausage Meat**
(a) shall be fresh or preserved comminuted meat;
(b) may be enclosed in a casing;
(c) may be dipped in vinegar, smoked cooked or dried;
(d) **may contain**
(i) **animal fat,**
(ii) **filler,**
(iii) **beef tripe,**
(iv) liver,
(v) fresh or frozen beef and pork blood,
(vi) **sweetening agents,**
(vii) salt and spices,
(viii) seasoning, other than tomato,
(ix) lactic acid producing starter culture,
(x) meat binder,
(xi) beef and pork blood plasma,
(xii) in the case of preserved comminuted meat, smoke flavouring or artificial smoke flavouring if the principal display panel of the label carries, immediately preceding or following the common name, the statement "Smoke Flavouring Added" or "Artificial Smoke Flavouring Added," whichever is applicable,
(xiii) if cooked
(1) glucono delta lactone;
(2) partially defatted beef fatty tissue or partially defatted pork fatty tissue; and
(3) a dried skim milk product, obtained from skim milk by the reduction of its calcium content and a corresponding increase in its sodium content, in an amount not exceeding 3 per cent of the finished food,
(xiv) in the case of fresh uncooked sausage,
(A) **artificial maple flavour,** if the principal display panel of the label carries, immediately preceding or following the common name, the statement "Artificial Maple Flavour Added," and
(B) apple powder as a flavouring ingredient, if the principal display panel of the label carries, immediately preceding or following the common name, the statement "Apple Powder Added for Flavouring;"

(xv) in the case of dry sausage or dry sausage meat, glucono delta lactone; and
(xvi) in the case of longaniza, annatto in such amount as will result in the finished product containing not more than 0.1 per cent annatto, if annatto is shown, by the word "annatto," in the list of ingredients on the label;

B.14.037. [S].
Headcheese
(a) shall be comminuted cooked meat or comminuted cooked preserved meat;
(b) shall not contain
(i) less than 50 per cent head meat, or
(ii) skin, other than that naturally adherent to any pork meat used;
(c) may contain scalps, snouts, beef tripe, salt, spices, seasoning or an added gelling agent; and
(d) may contain
(i) ascorbic acid or its sodium salt, or
(ii) erythorbic acid or its sodium salt.

Authors' Note: *Cases of food poisoning are becoming more and more common. The following regulations (B.14.061 and B.14.062) are examples of safeguards that formerly protected consumers to some extent from deadly bacteria, such as E.coli. You will note the sections that would have provided safeguards were removed as of March 1997 and we are at a loss to understand why. If no one is responsible for measuring or restricting the levels of bacteria in these products, who will be held responsible in the event of an outbreak of food poisoning?*

B.14.061. [S].
Edible Bone Meal or **Edible Bone Flour shall be the food prepared by grinding dry defatted bones,** obtained from animals healthy at the time of slaughter and shall contain
(a) not less than 85 per cent ash, as determined by official method FO-34.
* **(b) and (c) below were repealed as of March 19, 1997.**
(b) not more than 1,000 total aerobic bacteria per gram, as determined by official method MFO-1.
(c) no Escherichia coli, as determined by official method MFO-1.

B.14.062. [S].(1)
Gelatin or **Edible Gelatin**

(a) shall be the purified food obtained by the processing of skin, ligaments or bones of animals;
(b) shall contain not less than 82 per cent ash-free solids, when tested by official method FO-35.
(c) shall be free from objectionable taste and offensive odour when 2.5 grams thereof are dissolved in 100 millilitres of warm water;
* **(d) below was repealed as of March 19, 1997.**
(d) shall not contain
  (i) more than 5,000 bacteria per gram,
  (ii) more than 10 coliform bacteria per gram, or
  (iii) bacteria of the genus *Salmonella* as determined by official method MFO-13.
(e) shall not contain any residues of hydrogen peroxide where it has been used in the course of manufacture; and
(f) may contain
  (i) not more than 2.6 per cent ash on a dry basis,
  (ii) not more than 500 parts per million of sulphurous acid, including the salts thereof, calculated as sulphur dioxide, and
  (iii) **where intended for** use in the manufacture of **marshmallow, sodium hexametaphosphate or sodium lauryl sulphate.**

(2) **No person shall use, in** the course of manufacturing **gelatin or edible gelatin,**
(a) acidic or basic **compounds other than acetic acid, ammonium hydroxide, citric acid, fumaric acid, hydrochloric acid, lime magnesium hydroside, phosphoric acid, sodium carbonate, sodium hydroxide, sodium sulphide, sulphuric acid, sulphurous acid or tartaric acid;** or
(b) **filtering and clarifying agents** other than **activated carbon, alumina, aluminum sulphate, calcium phosphate, dibasic, cellulose, diatomaceous earth, perlite,** strongly acidic cation exchange resin in the hydrogen ion form or basic anion exchange resins in the chloride ion or free base ion forms.

B.14.063.
For the purposes of section B.14.064 to B.14.068, "stew meat" means meat that contains when raw not more than
- (a) 25 per cent fat, in the case of meat in meat ball stews; and
- (b) 20 per cent fat, in the case of meat in other stews.

B.14.064. [S].
Vegetable Stew with (naming the meat)
- (a) shall contain vegetables and the named meat in the following amounts, calculated as raw ingredients;
  - (i) 12 per cent or more stew meat, and
  - (ii) 38 per cent or more vegetables; and
- (b) may contain gravy, salt, seasoning and spices.

B.14.065. [S].
(naming the meat) Stew
- (a) shall contain vegetables and the named meat in the following amounts, calculated as raw ingredients:
  - (i) 20 per cent or more stew meat, and
  - (ii) 30 per cent or more vegetables; and
- (b) may contain gravy, salt, seasoning and spices.

B.14.070. [S].
Wieners and Beans or **Wieners with Beans** shall be the food prepared from dried beans and wieners, may contain sauce, seasoning, spices and a sweetening agent, and shall contain not less than 25 per cent wieners, as determined by official method FO-36.

B.14.071. [S].
Beans and Wieners or **Beans with Wieners** shall be the food prepared from dried beans and wieners, may contain sauce, seasoning, spices and a sweetening agent and shall contain not less than 10 per cent wieners, as determined by official method FO-36.

DIVISION 15

Adulteration of Food

Authors' Note: *The table below came as a shock to us. Perhaps we should be relieved that they have set limits on the amount of toxic substances permitted in food. We wonder if they are finding it easy to regulate and enforce these limits.*

B.15.001.

A food named in column III of an item of Table I to this Division is adulterated if the substance named in column I of that item is present therein or has been added thereto in an amount exceeding the amount, expressed in parts per million, shown in column II of that item for that food.

Substance	Tolerance ppm	Foods
1.Arsenic	(1) 3.5	(1) Fish protein
	(2) 1	(2) Edible bone meal
	(3) 0.1	(3) Fruit juice, fruit nectar, beverages when ready-to-serve and water in sealed containers other than mineral water or spring water
2. Fluoride	(1) 650	(1) Edible bone meal
	(2) 150	(2) Fish protein
3. Lead	(1) 10	(1) Edible bone meal
	(2) 1.5	(2) Tomato paste and tomato sauce
	(3) 0.5	(3) Fish protein and whole tomatoes
	(4) 0.2	(4) Fruit juice, fruit nectar, beverages when ready-to-serve and water in sealed containers other than mineral water or spring water
	(5) 0.15	(5) Evaporated milk, condensed milk and concentrated infant formula
	(6) 0.08	(6) Infant formula when ready-to-serve
4. Tin	(1) 250	(1) Canned foods

B.15.002.

(1) Subject to subsection (2) and (3), a food is adulterated if an agricultural chemical or any of its derivatives is present therein or has been added thereto, singly or in any combination, in an amount exceeding 0.1 parts per million, unless it is listed and used in accordance with the tables to Division 16.

(2) Subject to subsection (3), a food is exempt from paragraph 4(d) of the Act if the only agricultural chemicals that are present therein or have been added thereto are any of the following:
(a) a fertilizer;
(b) an adjuvant or a carrier of an agricultural chemical;
(c) an inorganic bromide salt;
(d) silicon dioxide;
(e) sulphur; or
(f) viable spores of *Bacillus thuringiensis* Berliner.

(3) A food named in column IV of an item of Table II to this Division is exempt from paragraph 4(d) of the Act if the agricultural chemicals named in columns I and II of that item are present therein or have been added thereto in an amount not exceeding the limit, expressed in parts per million, set out in column III of that item for that food.

B.15.003.
A food named in column IV of an item of Table III, to this Division is exempt from paragraph 4(d) of the Act **if the drug named in column I,** and analysed as being the substance named in column II, of that item is present in the food in an amount not exceeding the limit, expressed in parts per million, set out in column III of that item for that food.

Authors' Note: *For a list of chemicals and drugs that could end up in your food, see Tables of Agricultural Chemicals and Veterinary Drugs.*

DIVISION 16

Food Additives

B.16.001.
A quantitative statement of the amount of each additive present or directions for use that, if followed, will produce a food that will not contain such additives in excess of the maximum levels of use prescribed by these Regulations shall be shown, grouped together with the list of ingredients, of any substance or mixture of substances for use as a food additive.

Authors' Note: *We found it interesting that a food manufacturer who requests that a substance be added to the food additive tables, or a change be made to the existing food additive tables simply has to follow the instructions in B.16.002 (below) and within ninety days could have the Minister's approval (B.16.003). The evaluation of safety seems to be left up to the manufacturer, contrary to the public's assumption that chemicals are not added to food without years of onerous testing for safety by government or independent researchers.*

B.16.002.

A request that a food additive be added to or a change made **in the Tables following section B.16.100 shall be accompanied by a submission to the Minister** in a form, manner and content satisfactory to him **and shall include**

(a) **a description of the food additive,** including its chemical name and the name under which it is proposed to be sold, its method of manufacture, its chemical and physical properties, its composition and its specifications and, where that information is not available, a detailed explanation;

(b) a statement of **the amount of the food additive proposed for use,** and the purpose for which it is proposed, together with all directions, recommendations and suggestions for use;

(c) where necessary, in the opinion of the Director, **an acceptable method of analysis** suitable for regulatory purposes that will determine the amount of the food additive and of any substance resulting from the use of the food additive in the finished food;

(d) **data establishing that the food additive will have the intended** physical or other technical **effect;**

(e) **detailed reports of tests made to establish the safety of the food additive under the conditions of use** recommended;

(f) **data to indicate the residues that may remain in or upon the finished food** when the food additive is used in accordance with good manufacturing practice;

(g) **a proposed maximum limit for residues of the food additive in or upon the finished food;**

(h) **specimens of the labelling** proposed for the food additive; and

(i) **a sample of the food additive** in the form in which it is proposed to be used in foods, a sample of the active ingredient, and, on request a sample of food containing the food additive.

B.16.003.
The Minister shall, within ninety days after the filing of a submission in accordance with section B.16.002, **notify the person filing the submission whether or not it is his intention to recommend to the Governor-in-Council that the said food additive be so listed** and the detail of any listing to be recommended.

B.16.004.
No person shall use in or upon a food more than one of the following Class II preservatives:

(a) benzoic acid including salts thereof;
(b) methyl-p-hydroxybenzoate and propyl-p-hydroxybenzoate; including salts thereof; and
(c) sulphurous acid including salts thereof.

B.16.006
Paragraph (c) of section B.01.042 and paragraph (a) of section B.01.043 do not apply to spices, seasonings, flavouring preparations, essential oils, oleoresins and natural extractives.

Authors' Note: *Regulation B.16.006 is an example of Health Canada creating a law and then canceling it with another regulation. There are so many of these "subsection does not apply to" and "notwithstanding clauses" that we feel it gives food manufacturers almost* carte blanche *to manipulate the regulations to produce their food products any way they choose.*

B.16.007.
No person shall sell a food containing a food additive other than a food additive provided for in sections B.01.042, B.01.043 and B.25.062.

B.16.100.
No person shall sell any substance as a food additive unless the food additive is listed in one or more of the following Tables:

Authors' Note: See *Tables of Additives According to Function.*

DIVISION 17

Salt

B.17.001.(1) [S].

Salt, other than crude rock salt, shall be crystalline sodium chloride and may contain:

(a) **one or more of the following anticaking agents,**

  (i) **calcium aluminum silicate, calcium phosphate tribasic, calcium silicate, calcium stearate, magnesium carbonate, magnesium silicate, magnesium stearate, silicon dioxide and sodium aluminum silicate,** the total amount not to exceed 1 per cent and, in the case of fine grained salt, the total amount not to exceed 2 per cent,

  (ii) **propylene glycol** in an amount not exceeding 0.035 per cent, and

  (iii) **sodium ferrocyanide decahydrate** in an amount not exceeding 13 parts per million calculated as anhydrous sodium ferrocyanide;

(b) not more than

  (i) 1.4 per cent, **singly or in combination, of calcium sulphate or potassium chloride,**

  (ii) 13 parts per million sodium ferrocyanide when added as sodium ferrocyanide decahydrate in the production of dendritic crystals of salt,

  (iii) 10 parts per million of polyoxyethylene (20) sorbitan monooleate when used in the production of coarse crystal salt,

  (iv) 15 parts per million of sodium alginate when used in the production of coarse crystal salt, and

  (v) 0.1 per cent other ingredients; and

(c) notwithstanding paragraphs (a) and (b), the total level of sodium ferrocyanide decahydrate, whether added as an anticaking agent or as an adjuvant in the production of dendritic salt, shall not exceed 13 parts per million, calculated as anhydrous sodium ferrocyanide.

DIVISION 18

Sweetening Agents

B.18.001. [S].
Sugar

(a) shall be the food chemically known as sucrose; and
(b) shall contain not less than 99.8 per cent sucrose.

B.18.002. [S].
Liquid Sugar shall be the food obtained by dissolving sugar in water.

B.18.003. [S].
Invert Sugar shall be the food obtained by the partial or complete hydrolysis of sugar.

B.18.004. [S].
Liquid Invert Sugar shall be the food consisting of a solution of invert sugar in water.

B.18.006. [S].
Icing Sugar

(a) shall be powdered sugar; and
(b) may contain
   (i) **food colour,** and
   (ii) either not more than 5 per cent starch or an anticaking agent.

B.18.007. [S].
Brown Sugar, Yellow Sugar or **Golden Sugar**

(a) shall be the food obtained from the syrups originating in the sugar refining process:
(b) may contain not more than
   (i) 4.5 per cent moisture, and
   (ii) 3.5 per cent sulphated ash; and
(c) shall not contain less than 90 per cent sugar and invert sugar.

B.18.008. [S].
Refined Sugar Syrup, Refiners' Syrup or **Golden Syrup**

(a) shall be the food made from syrup originating in the sugar refining process;

(b) may be hydrolyzed; and
(c) may not contain more than
  (i) 35 per cent moisture, and
  (ii) 2.5 per cent sulphated ash.

B.18.009. [S].
Fancy Molasses
(a) shall be the syrupy food obtained by the evaporation and partial inversion of the clarified or unclarified sugar cane juice from which sugar has not been previously extracted;
(b) may contain sulphurous acid or its salts;
(c) shall not contain more than
  (i) 25 per cent moisture, and
  (ii) 3 per cent sulphated ash.

B.18.010. [S].
Table Molasses
(a) shall be the liquid food obtained in the process of manufacturing raw or refined sugar;
(b) may contain sulphurous acid or its salts;
(c) shall not contain more than
  (i) 25 per cent moisture, and
  (ii) 3 per cent sulphated ash.

B.18.016. [S].
Glucose or Glucose Syrup
(a) shall be the purified concentrated solution of nutritive saccharides obtained from the incomplete hydrolysis, by means of acid or enzymes, of starch or of the starch-containing substance;
(b) shall have a total solids content of not less than 70 per cent;
(c) shall have a sulphated ash content of not more than 1.0 per cent on a dry basis;
(d) shall have a reducing sugar content (dextrose equivalent) of not less than 20 per cent expressed as D-glucose on a dry basis; and
(e) may contain sulphurous acid or its salts.

B.18.017. [S].
Glucose Solids or **Dried Glucose Syrup**
(a) shall be glucose or glucose syrup from which the water has been partially removed;

(b) shall have a total solids content of not less than 93 per cent;
(c) shall have a sulphated ash content of not more than 1.0 per cent on a dry basis;
(d) shall have a reducing sugar content (dextrose equivalent) of not less than 20 per cent expressed as D-glucose, on a dry basis; and
(e) may contain sulphurous acid or its salts.

Authors' Note: *Corn syrup is a good example of a syrup (below) which we have found in some baby formulas. Any knowledgeable doctor will tell you a baby does not need most of these ingredients.*

B.18.018. [S].
(Naming the source of the glucose) Syrup
(a) shall be glucose;
(b) **may contain**
(i) **a sweetening agent,**
(ii) **a flavouring preparation,**
(iii) sorbic acid,
(iv) sulphurous acid or its salts,
(v) salt, and
(vi) water; and
(c) shall not contain more than
(i) 35 per cent moisture; and
(ii) 3 per cent ash.

B.18.025. [S].
Honey shall be the food produced by honey bees and derived from
(a) the nectar of blossoms,
(b) secretions of living plants, or
(c) secretions on living plants, and shall have
(d) a fluid, viscous or partly or wholly crystallized consistency;
(e) a diastase activity, determined after processing and blending, as represented by a diastase figure on the Gothe scale of not less than 8 where the hydroxy-methyl-furfural content is not more than 0.004 per cent; or
(f) a diastase activity, determined after processing and blending, as represented by a diastase figure on the Gothe scale of not less than 3 where the hydroxy-methyl-furfural content is not more than 0.0015 per cent.

DIVISION 19

Vinegar

B.19.001.
Vinegar shall be the liquid obtained by the acetous fermentation of an alcoholic liquid and shall contain not less than 4.1 per cent and not more than 12.3 per cent acetic acid.

B.19.002.
The percentage of acetic acid by volume contained in any vinegar described in Division 19 shall be shown on the principal display panel followed by the words "acetic acid."

B.19.003. [S].
Wine Vinegar shall be vinegar made from wine and may contain caramel.

B.19.004. [S].
Spirit Vinegar, Alcohol Vinegar, White Vinegar or **Grain Vinegar** shall be vinegar made from diluted distilled alcohol.

B.19.005. [S].
Malt Vinegar shall be vinegar made from an infusion of malt undistilled prior to acetous fermentation, and may contain other cereals or caramel, shall be dextro-rotatory, and shall contain, in 100 millilitres measured at a temperature of 20°C, not less than

(a) 1.8 grams of solids; and
(b) 0.2 gram of ash.

B.19.006. [S].
Cider Vinegar or **Apple Vinegar** shall be vinegar made from the liquid expressed from whole apples, apple parts or apple culls and may contain caramel.

B.19.007. [S].
Blended Vinegar shall be a combination of two or more varieties of vinegar of which spirit vinegar shall contribute not more than 55 per cent of the total acetic acid.

DIVISION 20

Tea

B.20.001. [S].
Tea shall be the dried leaves and buds of *Thea sinensis* (L.) Sims prepared by the usual trade processes.

B.20.002. [S].
Black Tea shall be black tea or a blend of two or more black teas and shall contain, on the dry basis, not less than 30 per cent water-soluble extractive, as determined by official method FO-37, and not less than 4 per cent and not more than 7 per cent total ash.

B.20.004. [S].
Green Tea shall contain, on the dry basis, not less than 33 per cent water-soluble extractive, as determined by official method FO-37, not less than 4 per cent and not more than 7 per cent total ash.

B.20.005. [S].
Decaffeinated (indicating the type of tea)
- (a) shall be tea of the type indicated, from which caffeine has been removed and that, as a result of the removal, contains not more than 0.4 per cent caffeine; and
- (b) **may have been decaffeinated by means of extraction solvents set out in Table XV to Division 16.**

DIVISION 21

Marine and Fresh Water Animal Products

B.21.001.
The foods referred to in this Division are included in the term marine and fresh water animal products.

B.21.002.
In this Division
"filler" means
- (a) flour or meal prepared from grain or potato, but not from a legume;

(b) processed wheat flour containing not less than the equivalent of 80 per cent dextrose, as determined by official method FO-32;
(c) bread, biscuit or bakery products, but not those containing or made with a legume;
(d) milk powder, skim milk powder, buttermilk powder or whey powder; and
(e) starch;

"Marine and fresh water animal" includes
(a) fish;
(b) crustaceans, molluscs, other marine invertebrates,
(c) marine mammals, and
(d) frogs.

B.21.003. [S].
Fish shall be the clean, dressed edible portion of fish, with or without salt or seasoning, and may
(a) **in the case of frozen fillets,** contain ascorbic acid or its sodium salt or erythorbic acid or its sodium salt and
(i) **sodium tripolyphosphate, sodium hexametaphosphate or a combination of sodium tripolyphosphate, sodium acid pyrophosphate and sodium pyrophosphate tetrabasic, or**
(ii) **a mixture of sodium hexametaphosphate and sodium carbonate;**
(b) **if frozen,** have a glaze consisting of water, **acetylated monoglycerides, calcium chloride, sodium alginate, sodium carboxymethyl cellulose, sodium phosphate (dibasic), corn syrup, dextrose, glucose, glucose solids,** ascorbic acid or its sodium salt or erythorbic acid or its sodium salt; and
(c) **if frozen minced, contain sodium tripolyphosphate, sodium hexametaphosphate,** ascorbic acid or its sodium salt, erythorbic acid or its sodium salt, **or a combination of sodium tripolyphosphate, sodium acid pyrophosphate and sodium pyrophosphate tetrabasic.**

B.21.004. [S].
In this Division, meat shall be the clean, dressed flesh of crustaceans, molluscs, other marine invertebrates and marine mammals, whether minced or not, with or without salt or seasoning, and in the case of frozen lobster, frozen crab, frozen shrimp and frozen clams, **may**

contain sodium tripolyphosphate or sodium hexametaphosphate or a combination of sodium hexametaphosphate and sodium carbonate or a combination of sodium tripolyphosphate, sodium acid pyrophosphate and sodium pyrophosphate tetrabasic.

B.21.006. [S].

Prepared fish or prepared meat shall be the whole or comminuted food prepared from fresh or preserved fish or meat respectively, may be canned or cooked, and may,

(a) in the case of **lobster paste** or fish roe (caviar), **contain food colour;**

(b) in the case of canned shellfish, canned spring mackerel and frozen cooked shrimp, contain citric acid or lemon juice;

(c) in the case of **fish paste, contain filler, fish binder, monoglycerides or mono- and diglycerides;**

(d) **in the case of canned salmon, tuna, lobster, crab-meat and shrimp, contain calcium disodium ethylenediaminetetraacetate (calcium disodium EDTA) and aluminum sulphate;**

(e) in the case of canned tuna, contain ascorbic acid;

(f) in the case of **canned sea foods,** contain **sodium hexametaphosphate and sodium acid pyrophosphate;**

(g) **contain liquid smoke flavour or liquid smoke flavour concentrate,** if the presence of such flavour is declared on the principal display panel of the label;

(h) contain edible oil, vegetable broth and tomato sauce or purée;

(i) **contain a gelling agent** if the principal display panel carries the word "jellied" as an integral part of the common name;

(j) contain salt;

(k) in the case of canned snails, **canned sea snails and canned clams, contain calcium disodium ethylenediamine tetra-acetate;**

(l) in the case of **canned flaked tuna, contain sodium sulphite;**

(m) in the case of lumpfish caviar, contain tragacanth gum;

(n) in the case of **a blend of prepared fish and prepared meat that has the appearance and taste** of the flesh of a marine or freshwater animal, **contain filler,** fish binder, whole egg, egg-white, egg-yolk, **food colour, gelling or stabilizing agents, texture-modifying agents,** natural and **artificial flavouring preparations, pH-adjusting agents, sweetener** and, in a

proportion not exceeding 2 per cent of the blend, a legume;
(o) in the case of crustaceans, contain potassium bisulphite, potassium metabisulphite, sodium bisulphite, sodium dithionite, sodium metabisulphite, sodium sulphite or sulphurous acid; and
(p) in the case of frozen crustaceans and molluscs, contain calcium oxide and sodium hydroxide.

B.21.007. [S].
Fish binder for use in or upon prepared fish or prepared meat **shall be filler with any combination of salt, sugar, dextrose, glucose, spices and other seasonings.**

B.21.021. [S].
Preserved fish or **preserved meat** shall be cooked or uncooked fish or meat that is dried, salted, pickled, cured or smoked and may contain **Class I preservatives, dextrose, glucose, spices, sugar** and vinegar, and
(a) dried fish that has been smoked or salted, and cold processed smoked and salted fish paste may contain sorbic acid or its salts;
(b) smoked fish may contain **food colour;** and
(c) **packaged fish and meat products** that are marinated or otherwise cold-processed **may contain sauderswood (sandalwood),** benzoic acid or its salts, **methyl-p-hydroxy benzoate and propyl-p-hydroxy benzoate.**
(d) **salted anchovy, salted scad and salted shrimp may contain erythrosine** in such amount as will result in the finished product containing not more than 125 parts per million of erythrosine.

B.21.024.
Notwithstanding section B.21.020 lobster paste shall not contain more than 2 per cent filler or fish binder.

Authors' Note: *It is beyond our comprehension why sections c(ii) and (iii) below, which provided some safeguards for food, were removed.*

B.21.027. (S).
Fish Protein
(a) shall be the food prepared by
(i) extracting water, fat and other soluble components

through the use of isopropyl alcohol from fresh whole edible fish of the order *Clupeiformes*, families *Clupeidae* and *Osmeridae* and the order *Gadiformes*, family *Gadidae*, or from trimmings resulting from the filleting of such fish when eviscerated, and

(ii) drying and grinding the protein concentrate resulting from the operation described in subparagraph (1);

(b) may contain a pH-adjusting agent; and

(c) shall not contain

(i) less than 75 per cent protein, which protein shall be at least equivalent to casein in protein quality as determined by official method.

* **(ii) and (iii) below were repealed as of March 19, 1997.**

(ii) more than 10,000 total aerobic bacteria per gram, as determined by official method MFO-16.

(iii) Escherichia coli, as determined by official method MFO-16.

DIVISION 22

Poultry, Poultry Meat, Their Preparations and Products

B.22.001. [S].
Poultry shall be any bird that is commonly used as food.

B.22.002. [S].
Poultry meat shall be the clean, dressed flesh, including the heart and gizzard of eviscerated poultry that is healthy at the time of slaughter.

B.22.003. [S].
Poultry meat by-product shall be the clean parts of poultry other than poultry meat commonly used as food and includes liver and skin, but excludes the esophagus, feet and head.

B.22.004. [S].
Giblets shall be the heart, liver and gizzard of poultry.

B.22.005.
Poultry meat, poultry meat by-products or preparations thereof are adulterated if any of the following substances or any substance in the following classes is present therein or has been added thereto:

(a) any organ or portion of poultry that is not commonly sold as food;
(b) **preservatives, other than those provided for in this Division;** and
(c) **colour, other than caramel.**

B.22.006. [S].
Prepared poultry meat or **a prepared poultry meat by-product** shall be any poultry meat or any poultry meat by-product, respectively, whether comminuted or not, to which has been added any ingredient permitted by these Regulations or that has been preserved, placed in a hermetically-sealed container or cooked, and may contain
(a) where a minimum total protein content or minimum meat protein requirement is prescribed in this Division, phosphate salts that do not when calculated as sodium phosphate, dibasic, exceed the maximum level provided therefor in Table XII to section B.16.100 and that are **one or more of the following** phosphate salts, namely,
(i) **sodium acid pyrophosphate,**
(ii) **sodium hexametaphosphate,**
(iii) **sodium phosphate, dibasic,**
(iv) **sodium phosphate, monobasic,**
(v) **sodium pyrophosphate, tetrabasic,**
(vi) **sodium tripolyphosphate,**
(vii) **potassium phosphate, monobasic,**
(viii) **potassium phosphate, dibasic, and**
(ix) **potassium pyrophosphate, tetrabasic; and**
(b) in the case of dried, cooked poultry meat, **a Class IV preservative.**

B.22.008.
In this Division, “filler” means any vegetable material (except tomato or beetroot), milk, egg, yeast, or any derivative or combination thereof that is acceptable as food.

B.22.010.
Powdered hydrogenated cotton seed oil in an amount not greater than 0.25 per cent of the product may be applied as a release agent to the surface of poultry meat, poultry meat by-product, prepared poultry meat, prepared poultry meat by-product, extended poultry product and simulated poultry product.

B.22.011. [S].
Solid cut poultry meat shall be
(a) a whole cut of poultry meat; or
(b) a product consisting of pieces of poultry meat of which at least 80 per cent weigh at least 25 g each.

B.22.012.
(1) **No person shall sell solid cut poultry meat** to which phosphate salts or water has been added unless
(a) that meat
(i) where cooked, contains a meat protein content of not less than 12 per cent, and
(ii) where uncooked, contains a meat protein content of not less than 10 per cent; and
(b) **that meat contains** phosphate salts that do not, when calculated as sodium phosphate, dibasic, exceed the maximum level provided therefor in Table XII to section B.16.100 and that are **one or more of the following** phosphate salts, namely,
(i) **sodium acid pyrophosphate,**
(ii) **sodium hexametaphosphate,**
(iii) **sodium phosphate, dibasic,**
(iv) **sodium phosphate, monobasic,**
(v) **sodium pyrophosphate, tetrabasic,**
(vi) **sodium tripolyphosphate,**
(vii) **potassium phosphate, monobasic,**
(viii) **potassium phosphate, dibasic, and**
(ix) **potassium pyrophosphate, tetrabasic.**

B.22.017. [S].
Vegetable Stew with (naming the poultry meat)
(a) shall contain vegetables and the named poultry meat in the following amounts:
(i) if uncooked, 12 per cent or more of the named stew poultry meat,
(ii) if cooked, 6 per cent or more of the named stew poultry meat,
(iii) 38 per cent or more vegetables; and
(b) may contain gravy, salt, seasoning and spices.

B.22.018. [S].
(naming the poultry meat) Stew
(a) shall contain vegetables and the named poultry meat in the following amounts:
(i) if uncooked, 20 per cent or more of the named stew poultry meat,
(ii) if cooked, 10 per cent or more of the named stew poultry meat,
(iii) 30 per cent or more vegetables; and
(b) may contain gravy, salt seasonings and spices.

B.22.021. [S].
Preserved poultry meat or **preserved poultry meat by-product** shall be cooked or uncooked poultry meat or poultry meat by-product that is cured or smoked and may contain
(a) **Class I preservatives;**
(b) **liquid smoke flavour, liquid smoke flavour concentrate** or spices:
(c) sweetening agents;
(d) vinegar; and
(e) in the case of cured poultry or poultry meat prepared by means of injection or cover solution, **disodium phosphate, monosodium phosphate, sodium hexametaphosphate, sodium tripolyphosphate, tetrasodium pyrophosphate and sodium acid pyrophosphate,** in such amount calculated as disodium phosphate, as will result in the finished product containing not more than 0.5 per cent added phosphate.

B.22.022. [S].
Canned (naming the poultry) shall be prepared from poultry meat and may contain
(a) those bones or pieces of bones attached to the portion of the poultry meat that is being canned;
(b) broth;
(c) salt;
(d) seasoning;
(e) **gelling agents;** and
(f) small amounts of fat.

B.22.024.
Where a gelling agent has been added to canned poultry, a statement to the effect that a gelling agent has been added shall be shown on the principal display panel or the word "jellied" shall be shown as an integral part of the common name of the food.

B.22.025. [S].
Boneless (naming the poultry) shall be canned poultry meat from which the bones and skin have been removed, shall contain not less than 50 per cent of the named poultry meat, as determined by official method FO-39 and may contain broth having a specific gravity of not less than 1.000 at a temperature of 50°C.

Egg Products

B.22.034. [S].
Liquid Whole Egg, Dried Whole Egg or **Frozen Whole Egg**
(a) shall be the product obtained by removing the shell from wholesome fresh eggs or wholesome stored eggs, and
  (i) in the case of dried whole egg, drying the product, or
  (ii) in the case of frozen whole egg, freezing the product; and
(b) **may**
  (i) **contain aluminum sulphate, pH-adjusting agents** and the colour beta-carotene,
  (ii) in the case of liquid whole egg destined for drying, contain yeast autolysate **and may be treated** with **hydrogen peroxide and catalase, glucose oxidase and catalase** or yeast and suitable glucose fermenting bacterial culture, or
  (iii) **in the case of dried whole egg, contain anticaking agents.**

B.22.035. [S].
Liquid Yolk, Dried Yolk or **Frozen Yolk**
(a) shall be the product obtained by removing the shell and egg-white from wholesome fresh eggs or wholesome stored eggs, and
  (i) in the case of dried yolk, drying the product, or
  (ii) in the case of frozen yolk, freezing the product, and
(b) may
  (i) contain aluminum sulphate, pH-adjusting agents and the colour beta-carotene,

(ii) **in the case of liquid yolk destined for drying, contain yeast autolysate and may be treated with hydrogen peroxide and catalase, glucose oxidase and catalase or yeast and suitable glucose fermenting bacterial culture, or**
(iii) **in the case of dried yolk, contain anticaking agents.**

B.22.036. [S].
Liquid Egg-White, (Liquid Albumen), Dried Egg-White, (Dried Albumen) or Frozen Egg-White (Frozen Albumen)
(a) shall be the product obtained by removing the shell and yolk from wholesome fresh eggs or wholesome stored eggs, and
(i) in the case of dried egg-white, drying the product, or
(ii) in the case of frozen egg-white, freezing the product; and
(b) **may**
(i) **contain whipping agents, aluminum sulphate and pH-adjusting agents,**
(ii) in the case of liquid egg-white destined for drying, contain yeast autolysate and **may be treated with hydrogen peroxide and catalase, glucose oxidase and catalase** or yeast and suitable glucose fermenting bacterial culture,
(iii) in the case of liquid egg-white and dried egg-white, contain **lipase or pancreatin,** or
(iv) in the case of dried egg-white, **contain anticaking agents.**

B.22.037. [S].
Liquid Whole Egg Mix, Dried Whole Egg Mix, Frozen Whole Egg Mix, Liquid Yolk Mix, Dried Yolk Mix or Frozen Yolk Mix
(a) shall be the product obtained by adding salt, sweetening agent or both to liquid whole egg, dried whole egg, frozen whole egg, liquid yolk, dried yolk, or frozen yolk; and
(b) may, in the case of dried whole egg mix or dried yolk mix, contain anticaking agents.

DIVISION 23

Food Packaging Materials

Authors' Note: *It appears to us that B.23.003 was written for the express purpose of cancelling out B.23.001 and B.23.002. What do you think?*

B.23.001.
No person shall sell any food in a package that may yield to its contents any substance that may be injurious to the health of a consumer of the food.

B.23.002.
Subject to section B.23.003 **no person shall sell any food in a package that has been manufactured from a polyvinyl chloride formulation containing an octyltin chemical.**

B.23.003.
A person may sell food, other than milk, skim milk, partly skimmed milk, sterilized milk, malt beverages and carbonated non-alcoholic beverage products, **in a package that has been manufactured from a polyvinyl chloride formulation containing any or all of the octyltin chemicals,** namely, di (n-octyl)tin S,S'-bis (isooctylmercaptoacetate), di (n-octyl)tin maleate polymer and (n-octyl)tin S,S',S"-tris (isooctylmercaptoacetate) if the proportion of such chemicals, either singly or in combination, does not exceed a total of 3 per cent of the resin, and the food in contact with the package contains not more than 1 part per million total octyltin.

B.23.004.
(1) Di (n-octyl)tin S,S'-bis (isooctylmercaptoacetate) shall be the octyltin chemical made from di (n-octyl)tin dichloride and shall contain 15.1 to 16.4 per cent of tin and 8.1 to 8.9 per cent of mercapto sulfur.

B.23.005.
Di (n-octyl)tin maleate polymer shall be the octyltin chemical made from di (n-octyl)tin dichloride and shall contain 25.2 to 26.6 per cent of tin.

B.23.006.

(1) (n-octyl)tin S,S',S"-tris (isooctylmercaptoacetate), being an octyltin chemical shall be made from (n-octyl)tin trichloride and shall contain 13.4 to 14.8 per cent of tin and 10.9 to 11.9 per cent of mercapto sulfur.

B.23.007.

No person shall sell a food in a package that may yield to its contents any amount of vinyl chloride, as determined by official method FO-40, in respect of that food.

B.23.008.

No person shall sell a food in a package that may yield to its contents any amount of acrylonitrile as determined by official method, FO-41, in respect of that food.

DIVISION 24

Foods for Special Dietary Use

B.24.001.

In this Division,

"formulated liquid diet" means a food that

- (a) is sold for consumption in liquid form, and
- (b) is sold or represented as a nutritionally complete diet for oral or tube feeding of a person described in paragraph (a) of the definition "food for special dietary use;"

"major change" means, in respect of a food that is represented for use in a very low energy diet, any change in any of the following, where the manufacturer's experience or generally accepted theory would predict an adverse effect on the levels or availability of nutrients in, the microbiological or chemical safety of or the safe use of the food:

- (a) an ingredient or the amount of an ingredient in the food;
- (b) the manufacturing process or the packaging of the food, or
- (c) the directions for the preparation and use of the food;

"very low energy diet" means a diet for weight reduction that provides less than 900 kilocalories per day when followed as directed.

B.24.004.
A carbohydrate-reduced food is a food
(a) that would, if it were not carbohydrate-reduced, derive at least 25 per cent of the calories contained in that food from its carbohydrate content; and
(b) that, when ready to serve,
(i) contains not more than 50 per cent of available carbohydrate normally found in that food when it is not carbohydrate-reduced as determined by an acceptable method, and
(ii) provides no more calories than would be provided if it were not carbohydrate-reduced.

B.24.005.
A sugar-free food is a carbohydrate-reduced food that, when ready to serve,
(a) contains not more than 0.25 per cent available carbohydrate as determined by an acceptable method; and
(b) provides, except in the case of chewing gum, not more than one calorie per 100 grams, or per 100 millilitres of that food.

B.24.006.
A calorie-reduced food is a food that, when ready to serve, provides not more than 50 per cent of the calories that would be normally provided in that food if it were not calorie-reduced.

B.24.007.
A low calorie food is a food that
(a) is a calorie-reduced food; and
(b) when ready to serve, provides not more than 15 calories per average serving and not more than 30 calories in a reasonable daily intake of that food as set out in Schedule K.

B.24.008.
A low sodium food is a food that, when ready to serve, contains
(a) not more than 50 per cent of the sodium that would normally be present in that food if it were not sodium reduced;
(b) not more than 40 milligrams of sodium per 100 grams of food, in the case of foods other than those described in paragraphs (c) and (d);

(c) not more than 80 milligrams of sodium per 100 grams of food, in the case of meat, fish and poultry products;
(d) not more than 50 milligrams of sodium per 100 grams of food, in the case of cheddar cheese; and
(e) except in the case of salt substitutes, no added salts of sodium.

Formulated Liquid Diets

B.24.100.
No person shall advertise a formulated liquid diet to the general public.

B.24.101.
No person shall sell a formulated liquid diet unless the food
(a) if sold ready to serve, or
(b) if not sold ready to serve, when diluted with water, milk, or water and milk, is a complete substitute for the total diet in meeting the nutritional requirements of a person.

DIVISION 25

Infant Foods, Infant Formula

B.25.001.
In this Division
"human milk substitute" means any food that is represented
(a) for use as a partial or total replacement for human milk and intended for consumption by infants, or
(b) for use as an ingredient in a food referred to in paragraph (a);

"infant" means a person who is under the age of one year;

"infant food" means a food that is represented for consumption by infants;

"junior (naming a food)" means the named food where it contains particles of a size to encourage chewing by infants, but may be readily swallowed by infants without chewing;

"major change" means, in respect of a human milk substitute, any change of an ingredient, the amount of an ingredient or the processing or packaging of the human milk substitute where the manufacturer's experience or generally accepted theory would predict an adverse effect on the levels or availability of nutrients in, or the microbiological or chemical safety of, the human milk substitute;

"new human milk substitute" means a human milk substitute that is
(a) manufactured for the first time,
(b) sold in Canada for the first time, or
(c) manufactured by a person who manufactures it for the first time;

"strained (naming a food)" means the named food where it is of generally uniform particle size that does not require and does not encourage chewing by infants before being swallowed;

Infant Foods

B.25.002.

No person shall sell or advertise for sale an infant food that is **set out in column I** of an item **of Table I to this Division and contains more than the amount of sodium set out in column II** of that item.

Sodium Content In Infant Foods

Food	Total Sodium in Grams per 100 grams of Food
1. Junior Desserts	0.10
2. Junior Meat, Junior Meat Dinners, Junior Dinners, Junior Breakfasts	0.25
3. Junior Vegetables, Junior Soups	0.20
4. Strained Desserts	0.05
5. Strained Meats, Strained Meat Dinners, Strained Dinners, Strained Breakfasts	0.15
6. Strained Vegetables, Strained Soups	0.10

B.25.003.

(1) Subject to subsection (2), **no person shall sell infant food that contains**
 (a) strained fruit,
 (b) fruit juice,
 (c) fruit drink, or
 (d) cereal,
if sodium chloride has been added to that food.

(2) **Subsection (1) does not apply to strained desserts containing any of the foods mentioned in paragraphs (1) (a) to (d).**

Human Milk Substitutes and Foods Containing Human Milk Substitutes

B.25.045.

The common name of a human milk substitute or a new human milk substitute shall be "infant formula."

B.25.046.

(1) **No person shall sell** or advertise for sale **a new human milk substitute unless the manufacturer,** at least **90 days before the sale** or advertisement, **notifies the Director in writing** of the intention to sell or advertise for sale the new human milk substitute.

(2) **The notification** referred to in subsection (1) shall be signed and **shall include,** in respect of the new human milk substitute, **the following information:**
 (a) **the name under which it will be sold** or advertised for sale;
 (b) **the name and the address** of the principal place of business of the manufacturer;
 (c) **the names and addresses of each establishment in which it is manufactured;**
 (d) **a list of all of its ingredients,** stated quantitatively;
 (e) **the specifications for nutrient,** microbiological and physical quality for the ingredients and for the new human milk substitute;
 (f) **details of quality control procedures** respecting the testing of the ingredients and of the new human milk substitute;

(g) **details of the manufacturing process** and quality control procedures used throughout the process;
(h) **the results of tests** carried out **to determine the expiration date** of the human milk substitute;
(i) **the evidence** relied on **to establish** that **the new human milk substitute is nutritionally adequate** to promote acceptable growth and development in infants when consumed in accordance with the directions for use;
(j) **a description** of the type **of packaging** to be used;
(k) **directions for use;**
(l) **the written text of all labels,** including package inserts, to be used in connection with the new human milk substitute; and
(m) **the name and title of the person** who signed the notification and the date of signature.

(3) **Notwithstanding subsection (1), a person may sell** or advertise for sale **a new human milk substitute if the manufacturer has notified the Director** pursuant to subsection (1) **and is informed in writing by the Director that the notification is satisfactory.**

Authors' Note: *Division 25 provides the regulations concerning "Human Milk Substitutes" which according to the* Act *is just another name for infant formula. Even though there are regulations governing the individual ingredients manufacturers may use to produce a new human milk substitute, this regulation permits manufacturers to choose from ingredients in Tables IV, V and X (listed in Food Additives According to Function) and to combine any of them to concoct a new product.*

Note the reference in regulation 25.046.(1) that stipulates the manufacturer must notify the Director of Health Canada in writing at least ninety days before he begins to sell his new milk substitute. This seems to be the time it takes the bureaucracy to process the paper work for the sale of the new product. It certainly is not nearly enough time for Health Canada to have the product tested for toxicity. There is nothing here to indicate manufacturers are required to provide results of animal testing for carcinogenicity, mutagenicity or teratogenicity. As parents, we consider animal testing more humane than using our babies as guinea pigs. We deem Regulations B.25.062(e) and (f) contributing factors to the escalation of childhood cancers in the last thirty-five years.

B.25.061.

(1) Subject to subsection (2), no person shall include on the label of a food any representation respecting the consumption of the food by an infant who is less than 6 months of age.

(2) Subsection (1) does not apply in respect of a human milk substitute or a new human milk substitute.

B.25.062.

(1) Subject to subsection (2), **no person shall sell a food that is labelled or advertised for consumption by infants if the food contains a food additive.**

(2) **Subsection (1) does not apply to**
- (a) **bakery products that are labelled or advertised for consumption by infants;**
- (b) ascorbic acid used in cereals containing banana that are labelled or advertised for consumption by infants:
- (c) soyabean lecithin used in rice cereal labelled or advertised for consumption by infants;
- (d) citric acid used in foods that are labelled or advertised for consumption by infants;
- (e) **infant formula that contains the food additives set out in Tables IV and X to section B.16.100 for use in infant formula; or**
- (f) **infant formula that contains ingredients manufactured with food additives set out in Table V to section B.16.100.**

Authors' Note: *If this regulation had ended with subsection (1), it would have been fine. However, subsection (2) pretty much negates what is said in the first paragraph. For example, aside from the additives that can be used in bakery products, take a close look at (e) and (f) which state that additives from Tables IV, V and X (listed in Tables of Additives According to Function) may be used in infant formula. We counted the chemicals in these Tables as well as the chemicals found in various ingredients used to produce infant formula and quit after passing the one hundred mark. This gives a considerable range of chemicals for manufacturers to choose from in their endeavour to surpass what is provided free by Mother Nature.*

DIVISION 26

Food Irradiation

Authors' Note: *Regulations governing irradiation are located in Division 1 and Division 26 – at opposite ends of the* Act. *Division 1 showed us the Radura irradiation symbol (a stylized flower) and gives information about irradiation labelling laws. Division 26 explains how irradiation is to be regulated and controlled. It would appear from these regulations that everything is in place to make it possible to add to the list of foods that may be irradiated.*

B.26.001.
In this Division, **"ionizing radiation"** means

- (a) **gamma-radiation from a Cobalt-60 or Cesium-137 source,**
- (b) X-rays generated from a machine source operated at or below an energy level of 5 MeV, and
- (c) electrons generated from a machine source operated at or below an energy level of 10 MeV;

"irradiation" means treatment with ionizing radiation.

B.26.002. This Division does not apply to foods exposed to ionizing radiation from a measuring instrument used to determine weight, estimate bulk solids, measure the total solids in liquids or perform other inspection procedures.

B.26.003.

- (1) Subject to subsection (2), **no person shall sell a food that has been irradiated.**
- (2) **A food set out in column I of an item of the table to this Division that has been irradiated may be sold if**
  - (a) **the food was irradiated from a source set out in column II** of that item for the purpose set out in column III of that item; and
  - (b) **the dose of ionizing radiation absorbed by the food is within the permitted absorbed dose set out in column IV of that item.**

Records

B.26.004.

(1) A manufacturer who sells a food that has been irradiated shall keep on his premises, for at least two years after the date of the irradiation, a record containing the following information:
   (a) the food irradiated and the quantity and lot numbers of the food;
   (b) the purpose of the irradiation;
   (c) the date of the irradiation;
   (d) the dose of ionizing radiation absorbed by the food;
   (e) the source of the ionizing radiation; and
   (f) a statement indicating whether the food was irradiated prior to the irradiation by the manufacturer and, if so, the information referred to in paragraphs (a) to (e) in respect of that prior irradiation.

(2) Every person who imports a food that is intended for sale in Canada that has been irradiated shall keep on his premises a record of the information referred to in subsection (1) for at least two years after the date of importation.

Changes to the Table

B.26.005. A request that a food be added or a change made to the table to this Division **shall be accompanied by a submission to the Director containing the following information:**
   (a) **the purpose and details of the proposed irradiation,** including the source of ionizing radiation and the proposed frequency of and minimum and maximum dose of ionizing radiation;
   (b) **data indicating that the minimum dose of ionizing radiation** proposed to be used accomplishes the intended purpose of the irradiation and the maximum dose of ionizing radiation proposed does not exceed the amount required to accomplish the purpose of the irradiation;
   (c) information on the nature of the dosimeter used, the frequency of the dosimetry on the food and data pertaining to the dosimetry and phantoms used to assure that the dosimetry readings reflect the dose absorbed by the food during irradiation;

(d) **data indicating the effects,** if any, **on the nutritional quality of the food, raw and ready-to-serve,** under the proposed conditions of irradiation and **any other processes that are combined with the irradiation;**
(e) data establishing that the **irradiated food has not been significantly altered** in chemical, physical or microbiological characteristics to render the food unfit for human consumption;
(f) where the Director so requests, data establishing that the proposed irradiation is safe under the conditions proposed for the irradiation;
(g) the recommended conditions of storage and shipment of the irradiated food including the time, temperature and packaging and a comparison of the recommended conditions for the same food that has not been irradiated;
(h) details of any other processes to be applied to the food prior to or after the proposed irradiation; and
(i) such other data as the Director may require to establish that consumers and purchasers of the irradiated food will not be deceived or misled as to the character, value, composition, merit or safety of the irradiated food.

Food	Permitted Sources of Ionizing Radiation	Purpose of Treatment	Permitted Absorbed Dose
1. Potatoes	Cobalt-60	To inhibit sprouting during storage	0.15 kGy max.
2. Onions	Cobalt-60	To inhibit sprouting during storage	0.15 kGy max.
3. Wheat, Flour, Whole Wheat Flour	3. Cobalt-60	To control insect infestation in stored food	0.75 kGy max.
4. Whole or ground spices and dehydrated seasoning preparations	4. Cobalt-60, Cesium-137, or electrons from machine sources (3 MeV max.)	To reduce microbial load	10.00 kGy max. total overall average dose

TABLES OF ADDITIVES ACCORDING TO FUNCTION

B.16.100.
No person shall sell any substance as a food additive unless the food additive is listed in one or more of the following Tables.

Authors' Note: *Throughout the* Food and Drugs Act, *you will find references such as "may contain anticaking agents, firming agents, sequestering agents, Class IV preservatives." The following tables list the additives according to their functions. For example, if a manufacturer requires an anticaking agent for his product, he chooses an additive from Table I, and so on.*

TABLE I

Anticaking Agents

Calcium Aluminum Silicate	Magnesium Silicate
Calcium Phosphate, tribasic	Magnesium Stearate
Calcium Silicate	Microcrystalline Cellulose
Calcium Stearate	Propylene Glycol
Cellulose	Silicon Dioxide
Magnesium Carbonate	Sodium Aluminum Silicate
Magnesium Oxide	Sodium Ferrocyanide, decahydrate

TABLE II

Bleaching, Maturing & Dough Conditioning Agents

Acetone Peroxide	Chlorine
Ammonium Persulphate	Chlorine Dioxide
Ascorbic Acid	l-Cysteine Hydrochloride
Azodicarbonamide	Potassium Iodate
Benzoyl Peroxide	Potassium Persulphate
Calcium Iodate	Sodium Stearoyl-2-Lactylate
Calcium Peroxide	Sodium Stearyl Fumarate
Calcium Stearoyl-2-Lactylate	Sodium Sulphite

TABLE III

Colouring Agents

Aluminum Metal	Titanium Dioxide
Alkanet	Turmeric
Annatto	Xanthophyll
Anthocyanins	ß-Apo-8'-carotenal
Beet Red	Ethyl ß-apo-8'-carotenoate
Canthaxanthin	Caramel
Carbon Black	Allura Red
Carotene	Amaranth
Charcoal	Erythrosine
Chlorophyll	Indigotine
Cochineal	Sunset Yellow FCF
Iron Oxide	Tartrazine
Orchil	Brilliant Blue FCF
Paprika	Fast Green FCF
Riboflavin	Citrus Red No. 2
Saffron	Ponceau SX
Saunderswood	Gold
Silver Metal	

TABLE IV

Emulsifying, Gelling, Stabilizing & Thickening Agents

Acacia Gum
Acetylated Monoglycerides
Acetylated Tartaric Acid Esters of Mono- and Diglycerides
Agar
Algin
Alginic Acid
Ammonium Alginate
Ammonium Carrageenan
Ammonium Furcelleran
Ammonium Salt of Phosphorylated Glyceride
Arabinogalactan
Baker's Yeast Glycan
Calcium Alginate
Calcium Carbonate
Calcium Carrageenan
Calcium Citrate
Calcium Furcelleran
Calcium Gluconate
Calcium Glycerophosphate
Calcium Hypophosphite
Calcium Phosphate, dibasic
Calcium Phosphate, tribasic
Calcium Sulphate
Calcium Tartrate
Carboxymethyl Cellulose
Carob Bean Gum
Carrageenan
Cellulose Gum
Furcelleran
Gelatin
Gellan Gum
Guar Gum
Gum Arabic
Hydroxylated Lecithin
Hydroxypropyl Cellulose
Hydroxypropyl Methylcellulose
Irish Moss Gelose
Karaya Gum
Lactylated Mono- and Diglycerides
Lactylic Esters of Fatty Acids
Lecithin
Locust Bean Gum
Magnesium chloride
Methylcellulose
Methyl Ethyl Cellulose
Monoglycerides
Mono- and Diglycerides
Monosodium Salts of Phosphorylated Mono- and Diglycerides
Oat Gum
Pectin
Polyglycerol Esters of Fatty Acids
Polyglycerol Esters of Interesterified Castor Oil Fatty Acids
Polyoxyethylene (20) Sorbitan Monooleate; Polysorbate 80
Polyoxyethylene (20) Sorbitan Monostearate; Polysorbate 60
Polyoxyethylene (20) Sorbitan Tristearate; Polysorbate 65
Polyoxyethylene (8) Stearate
Potassium Alginate
Potassium Carrageenan
Potassium Chloride
Potassium Citrate
Potassium Furcelleran
Potassium Phosphate, dibasic
Propylene Glycol Alginate
Propylene Glycol Ether of Methylcellulose
Propylene Glycol mono Fatty Acid Esters
Sodium Acid Pyrophosphate
Sodium Alginate
Sodium Aluminum Phosphate
Sodium Carboxymethyl Cellulose
Sodium Carrageenan
Sodium Cellulose Glycolate
Sodium Citrate
Sodium Furcelleran
Sodium Gluconate
Sodium Hexametaphosphate
Sodium Phosphate, dibasic
Sodium Phosphate, monobasic
Sodium Phosphate, tribasic
Sodium Potassium Tartrate
Sodium Pyrophospate, tetrabasic
Sodium Stearoyl-2-Lactylate
Sodium Tartrate
Sodium Tripolyphosphate
Sorbitan Monostearate
Sorbitan Trioleate
Sorbitan Tristearate
Stearyl Monoglyceridyl Citrate
Sucrose Esters of Fatty Acids
Tannic Acid
Tragacanth Gum
Xanthan Gum

Food Additives That May Be Used As Food Enzymes

Authors' Note: *According to Health Canada, no additives* per se *have been genetically engineered but several enzymes used as additives have been produced by genetically modified organisms (GMOs). We have inserted an asterisk next to each enzyme that was identified to us by the Bureau of Microbial Hazards at Health Canada as having a genetically engineered source.*

Paul Mayers, Acting Director of the Microbial Hazards Division, explained to us an enzyme is an enzyme regardless of the source from which it was taken. Just because the source from which it was taken is genetically engineered, doesn't mean the enzyme itself is genetically altered. Our concern with this explanation is "Who makes that determination and on what basis do they make the decision that the enzyme has not been genetically altered?"

A biochemist explained to us that the amino acid sequences of the enzyme from the unmodified strain and the amino acid sequences from the genetically modified strain should be compared. If their chemical structure has been altered even slightly they may still have the same biological activity. In other words, even though they are made the same way and may still function the same way, if there is even a slight difference in their chemical structure, they cannot be considered identical to the unengineered enzyme. If biological activity is all that is being measured by the Microbial Hazards Division, they can still claim the enzymes are identical. An enzyme from a GMO source that has a slight difference in its chemical structure could have unexpected side-effects.

Another major concern is the possibility of impurities in GMOs. It is generally understood that it is technically impossible to isolate an enzyme that is 100 per cent pure whether it is from a GMO source or not. We wonder whether bio-engineers have developed ways, and are consistently using them, to rule out any possible contaminants. Obviously, the technology of genetically modifying organisms needs much more independent research.

As GMOs are not required to be identified on a label, it is impossible for the consumer to know which source the enzyme was taken from. GMO sources are cheaper and we suspect that they are the most frequently used. Eventually, the more costly, natural sources of enzymes may go the way of the dinosaur!

TABLE V

Food Enzymes

Amylase*
Amylase (maltogenic)*
Bovine Rennet
Bromelain
Catalase
Cellulase
Chymosin A*
Chymosin B*
Ficin
Glucoamylase
(Amyloglucosidase; Maltase)
Glucanase
Glucose Oxidase
Glucose Isomerase
Hemicellulase
Invertase
Lactase
Lipase*
Lipoxidase
Milk coagulating enzyme
Pancreatin
Papain
Pectinase
Pentosanase
Pepsin
Protease
Pullulanase
Rennet
Trypsin

TABLE VI

Firming Agents

Aluminum Sulphate
Ammonium Aluminum Sulphate
Calcium Chloride
Calcium Citrate
Calcium Gluconate
Calcium Lactate
Calcium Phosphate, dibasic
Calcium Phosphate, monobasic
Calcium Sulphate
Potassium Aluminum Sulphate
Sodium Aluminum Sulphate

TABLE VII

Glazing and Polishing Agents

Acetylated Monoglycerides
Beeswax
Carnauba Wax
Candelilla Wax
Gum Arabic
Gum Benzoin
Magnesium Silicate
Mineral Oil
Petrolatum
Shellac
Spermaceti Wax
Zein

TABLE VIII

Miscellaneous Food Additives

Acetylated Monoglycerides
Aluminum Sulphate
Ammonium Persulphate
Beeswax
Benzoyl Peroxide
Brominated vegetable oil
n-Butane
Caffeine
Caffeine Citrate
Calcium Carbonate
Calcium Lactate
Calcium Oxide
Calcium Phosphate, dibasic
Calcium Phosphate, tribasic
Calcium Silicate
Calcium Stearate
Calcium Stearoyl-2-Lactylate
Calcium Sulphate
Carbon Dioxide
Castor Oil
Cellulose, Microcrystalline
Chloropentafluoroethane
Citric Acid
Copper Gluconate
Dimethylpolysiloxane Formulations
Dioctylsodium Sulfo-succinate
Ethoxyquin
Ethylene Oxide
Ferrous Gluconate
Glucono Delta Lactone
Glycerol
Hydrogen Peroxide
Isobutane
Lactylic Esters of Fatty Acids
Lanolin
Lecithin
L-Leucine
Magnesium Aluminum Silicate
Magnesium Carbonate
Magnesium Chloride
Magnesium Silicate
Magnesium Stearate
Magnesium Sulphate
Methyl Ethyl Cellulose
Microcrystalline Cellulose
Mineral Oil
Monoacetin
Mono- and diglycerides
Monoglycerides
Nitrogen
Nitrous Oxide
Octafluorocycloubutane
Oxystearin
Ozone
Pancreas Extract
Paraffin Wax
Petrolatum
Polyethylene Glycol
Polydextrose
Polyvinylpyrrolidone
Potassium Aluminum Sulphate
Potassium Stearate
Propane
Propylene Glycol
Quillaia Extract
Saponin
Sodium Acid Pyrophosphate
Sodium Aluminum Sulphate
Sodium Bicarbonate
Sodium Carbonate
Sodium Carboxymethyl cellulose
Sodium Citrate
Sodium Ferrocyanide, decahydrate
Sodium Hexametaphosphate
Sodium Hydroxide
Sodium Lauryl Sulphate
Sodium Methyl Sulphate
Sodium Potassium Copper Chlorophyllin
Sodium Phosphate, dibasic
Sodium Pyrophosphate, tetrabasic
Sodium Silicate
Sodium Stearate
Sodium Stearoyl-2-Lactylate
Sodium Sulphate
Sodium Sulphite
Sodium Thiosulphate
Sodium Tripolyphosphate
Stannous Chloride
Stearic Acid
Sucrose Acetate Isobutyrate
Sulphuric Acid
Talc
Tannic Acid
Triacetin
Triethyl Citrate

TABLE IX

Sweeteners

- Acesulfame Potassium
- Aspartame
- Aspartame, encapsulated to prevent degradation during baking
- Isomalt
- Lactitol
- Maltitol
- Maltitol Syrup
- Mannitol
- Sorbitol
- Sucralose
- Thaumatin
- Xylitol

TABLE X

pH Adjusting Agents, Acid-Reacting Materials & Water-Correcting Agents

- Acetic Acid
- Adipic Acid
- Ammonium Aluminum Sulphate
- Ammonium Bicarbonate
- Ammonium Carbonate
- Ammonium Citrate, dibasic
- Ammonium Citrate, monobasic
- Ammonium Hydroxide
- Ammonium Phosphate, dibasic
- Ammonium Phosphate, monobasic
- Calcium Acetate
- Calcium Carbonate
- Calcium Chloride
- Calcium Citrate
- Calcium Fumarate
- Calcium Gluconate
- Calcium Hydroxide
- Calcium Lactate
- Calcium Oxide
- Calcium Phosphate, dibasic
- Calcium Phosphate, monobasic
- Calcium Phosphate, tribasic
- Calcium Sulphate
- Citric Acid
- Cream of Tartar
- Fumaric Acid
- Gluconic Acid
- Glucono Delta Lactone
- Hydrochloric Acid
- Lactic Acid
- Magnesium Carbonate
- Magnesium Citrate
- Magnesium Fumarate
- Magnesium Hydroxide
- Magnesium Oxide
- Magnesium Phosphate
- Magnesium Sulphate
- Malic Acid
- Manganese Sulphate
- Phosphoric Acid
- Potassium Acid Tartrate
- Potassium Aluminum Sulphate
- Potassium Bicarbonate
- Potassium Carbonate
- Potassium Chloride
- Potassium Citrate
- Potassium Fumarate
- Potassium Hydroxide
- Potassium Lactate
- Potassium Phosphate, dibasic
- Potassium Sulphate
- Potassium Tartrate
- Sodium Acetate
- Sodium Acid Pyrophosphate
- Sodium Acid Tartrate
- Sodium Aluminum Phosphate
- Sodium Aluminum Sulphate
- Sodium Bicarbonate
- Sodium Bisulphate
- Sodium Carbonate
- Sodium Citrate
- Sodium Fumarate
- Sodium Gluconate
- Sodium Hexametaphosphate
- Sodium Hydroxide
- Sodium Lactate
- Sodium Phosphate, dibasic
- Sodium Phosphate, monobasic
- Sodium Phosphate, tribasic
- Sodium Potassium Tartrate
- Sodium Pyrophosphate, tetrabasic
- Sodium Tripolyphosphate
- Sulphuric Acid
- Sulphurous Acid
- Tartaric Acid

TABLE XI

Class I Preservatives

Acetic Acid
Ascorbic Acid
Calcium Ascorbate
Erythorbic Acid
Iso-ascorbic Acid
Potassium Nitrate
Potassium Nitrite
Sodium Ascorbate
Sodium Erythorbate
Sodium Iso-ascorbate
Sodium Nitrate
Sodium Nitrite

Class II Preservatives

Benzoic Acid
Calcium Sorbate
Methyl-p-hydroxy Benzoate
Methyl Paraben
Potassium Benzoate
Potassium Bisulphite
Potassium Metabisulphite
Potassium Sorbate
Propyl-p-hydroxy Benzoate
Propyl Paraben
Sodium Benzoate
Sodium Bisulphite
Sodium Metabusulphite
Sodium Salt of Methyl-p-hydroxy Benzoic Acid
Sodium Salt of Propyl-p-hydroxy Benzoic Acid
Sodium Sorbate
Sodium Sulphite
Sodium Dithionite
Sorbic Acid
Sulphurous Acid

Class III Preservatives

Calcium Propionate
Calcium Sorbate
Natamycin
Potassium Sorbate
Propionic Acid
Sodium Diacetate
Sodium Propionate
Sodium Sorbate
Sorbic Acid

Class IV Preservatives

Ascorbic Acid
Ascorbyl Palmitate
Ascorbyl Stearate
Butylated Hydroxyanisole (BHA)
Butylated Hydroxytoluene (BHT)
Citric Acid
L-Cysteine Hydrochloride
Gum Guaiacum
Lecithin
Lecithin Citrate
Monoglyceride Citrate
Mono-isopropyl Citrate
Propyl Gallate
Tartaric Acid
Tocopherols

TABLE XII

Sequestering Agents

Ammonium Citrate, dibasic
Ammonium Citrate, monobasic
Calcium Citrate
Calcium Disodium Ethylenediaminetetraacetate
Calcium Disodium EDTA
Calcium Phosphate, monobasic
Calcium Phosphate, tribasic
Calcium Phytate
Citric Acid
Disodium Ethylenediaminetetraacetate
Disodium EDTA
Glycine
Phosphoric Acid
Potassium Phosphate, monobasic
Potassium Phosphate, dibasic
Potassium Pyrophosphate, tetrabasic
Sodium Acid Pyrophosphate
Sodium Citrate
Sodium Hexametaphosphate
Sodium Phosphate, dibasic
Sodium Phosphate, monobasic
Sodium Pyrophosphate, tetrabasic
Sodium Tripolyphosphate
Stearyl Citrate

TABLE XIII

Starch Modifying Agents

Acetic Anhydride
Adipic Acid
Aluminum Sulphate
Epichlorohydrin
Hydrochloric Acid
Hydrogen Peroxide
Magnesium Sulphate
Nitric Acid
Octenyl Succinic Anhydride
Peracetic Acid
Phosphorus Oxychloride
Potassium Permanganate
Propylene Oxide
Sodium Acetate
Sodium Bicarbonate
Sodium Carbonate
Sodium Chlorite
Sodium Hydroxide
Sodium Hypochlorite
Sodium Trimetaphosphate
Sodium Tripolyphosphate
Succinic Anhydride
Sulphuric Acid

TABLE XIV

Yeast Foods

Ammonium Chloride
Ammonium Phosphate, dibasic
Ammonium Phosphate, monobasic
Ammonium Sulphate
Calcium Carbonate
Calcium Chloride
Calcium Citrate
Calcium Lactate
Calcium Phosphate, dibasic
Calcium Phosphate, monobasic
Calcium Phosphate, tribasic
Calcium Sulphate
Ferrous Sulphate
Manganese Sulphate
Phosphoric Acid
Potassium Chloride
Potassium Phosphate, dibasic
Potassium Phosphate, monobasic
Sodium Sulphate
Zinc Sulphate

TABLE XV

Carrier or Extraction Solvents

Acetone
Benzyl Alcohol
1,3-Butylene Glycol
Carbon Dioxide
Castor Oil
Ethyl Acetate
Ethyl Alcohol (Ethanol)
Ethyl Alcohol denatured with methanol
Glycerol (Glycerine)
Glyceryl Diacetate
Glyceryl Triacetate (Triacetin)
Glyceryl Tributyrate (Tributyrin)
Hexane
Isopropyl Alcohol (Isopropanol)
Methyl Alcohol (Methanol)
Methyl Ethyl Ketone (2-Butanone)
Methylene Chloride (Dichloromethane)
Monoglycerides and Diglycerides
Monoglyceride Citrate
2-Nitropropane
1,2-Propylene Glycol (1,2-Propanediol)
Propylene Glycol Monoesters and Diesters of Fat-forming Fatty Acids
Triethyl Citrate

TABLES OF AGRICULTURAL CHEMICALS AND VETERINARY DRUGS

Authors' Note: *The Tables of Agricultural Chemicals and Veterinary Drugs are too lengthy to reproduce in full. In the* Act, *they include:*

1. *the common or trade name of the chemical or drug*
2. *chemical name of the substance*
3. *the maximum residue limit measured in parts per million (ppm)*
4. *the foods which are permitted to have residues of these chemicals or drugs in or upon them.*

Given our space restrictions, we have only listed the chemicals by their common or trade names and included a sample of the types of foods in which they could be found. Our primary focus in this book has been chemical food additives but we feel strongly that you have the right to know what could be in your food from the time the seed is planted to the time the finished product reaches the store shelf. These extensive lists show the number of chemicals that could actually end up in or upon the final product.

The term "agricultural chemicals" tends to downplay the fact that the majority of these chemicals are actually poisonous pesticides. For the most part, residues of agricultural chemicals and veterinary drugs are out-of-sight and therefore, out-of-mind. As a result, they are rarely taken into consideration by consumers when purchasing foods. Fortunately, many concerned farmers are reverting to organic methods. Their numbers will increase as consumers learn more about the disastrous state of the food supply and give organic farmers their support. There are many authors writing excellent books on the subject. We encourage you to look more deeply into this important area.

Agricultural Chemicals

Foods	Chemical
Cauliflower, Peppers, Lettuce, Beans, Potatoes, Cranberries, Cabbage	Acephate
Meat, Milk, Corn, Soybeans, Dry Beans	Alachlor (Lasso)
Potatoes	Aldicarb (Temik)
Butter, Cheese, Milk, Dairy Products, Meats	Aldrin and Dieldrin
Pears, Apples, Organ Meats of Cattle and Hogs	Amitraz
Most Berries, Celery, Onions, Squash, Potatoes, Tomatoes, Cucumbers	Anilazine (Dyrene)
Apples, Grapes, Most Berries, Citrus Fruits, Peaches, Pears, Tomatoes, Cucumbers	Azinphosmethyl (Guthion)
Most Berries, Apples, Citrus Fruits, Carrots, Tomatoes, Mushrooms, Cucumbers, Melons	Benomyl, Carbendazim and Thiophanatemethyl
Butter, Cheese, Milk and Meats	BHC Isomers, Except Lindane
Citrus fruits	Biphenyl
Apples	Bromophos
Cherries, Plums, Apples, Most Berries, Tomatoes, Peaches, Pears, Grapes	Captan
Most Fruits, Vegetables, Berries, Nuts, Grains	Carbaryl
Carrots, Strawberries, Potatoes, Peppers	Carbofuran (Furadan)
Carrots, Potatoes, Turnips, Onions, Peppers, Strawberries	Carbofuran Phenolic Metabolites
Butter, Cheese, Milk, Meat of Cattle, Goats, Hogs, Poultry and Sheep	Chlordane
Wheat	Chlormequat (Cycocel)
Strawberries, Watermelons, Honeydew melons, Potatoes, Peppers, Broccoli, Cucumbers	Chlorthaldimethyl (Dacthal)

Celery, Peanuts, Carrots, Tomatoes, Broccoli, Cucumbers, Beans, Onions, Melons	Chlorothalonil
Potatoes	Chlorpropham (CIPC)
Some Meat of Cattle, Citrus Fruits	Chlorpyrifos
Soyabeans, Lentils, Potatoes, Flax Seed	Clethodim
Almonds, Apples, Pears, Nectarines, Peaches, Milk and Some Meat of Cattle, Goats, Hogs, Horses and Sheep	Clofentezine
Soyabeans	Clomazone
Barley, Oats, Wheat, Flax, Strawberries	Clopyralid
Fresh Fruits and Vegetables	Copper Compounds (Inorganic)
Meat of Cattle, Goats, Horses, Hogs, Poultry and Sheep	Coumaphos (Co-Ral)
Mung Bean Sprouts	4-CPA
Apples, Citrus Fruits, Beans, Broccoli, Grapes, Pears, Tomatoes, Peaches, Strawberries, Asparagus	Cypermethrin (Cymbush, Ripcord)
Citrus Fruits	2,4-D
Tomatoes, Grapes, Pears, Peanuts, Plums Peaches, Apples, Cherries, Brussels Sprouts	Daminozide (Alar)
Fish, Eggs, Fresh Vegetables, Butter, Cheese, Milk, Meat, Meat By-Products and Fat of Cattle, Poultry, Hogs and Sheep	DDT
Apples, Apricots, Broccoli, Carrots, Cherries, Cauliflower, Celery, Grapes, Lettuce, Onions, Pears, Peppers, Plums, Spinach, Strawberries, Tomatoes, Citrus Fruits, Peaches, Beans, Cucumbers, Melons, Squash, Parsley	Diazinon
Apples, Cherries, Peaches, Plums, Prunes, Strawberries	Dichlone (Phygon)
Peaches, Raspberries, Strawberries, Celery, Grapes, Carrots, Onions, Tomatoes, Plums, Cucumbers, Rhubarb, Garlic	Dichloran (Botran)
Tomatoes and Certain Non-perishable Packaged Foods	Dichlorvos

Citrus Fruits, Melons, Most Berries, Nuts, Apples, Beans, Cherries, Grapes, Cucumbers, Peaches, Pears, Peas, Peppers, Soyabeans, Squash, Tomatoes, Butter, Cheese, Milk, Meat of Cattle, Goats, Hogs, Poultry and Sheep	Dicofol (Kelthane)
Apples, Broccoli, Cherries, Lettuce, Pears, Spinach, Citrus Fruits, Beans, Blueberries, Celery, Strawberries, Peas, Peppers, Tomatoes	Dimethoate
Strawberries	Diphenamid
Apples	Diphenylamine
Lentils	Diquat (Reglone)
Potatoes, Beans, Broccoli, Cabbage, Peas, Lettuce, Spinach, Tomatoes	Disulfoton (Disyston)
Citrus Fruits, Corn, Grapes, Potatoes, Pineapple, Wheat, Asparagus	Diuron
Apples, Pears, Strawberries, Cherries	Dodine (Cyprex)
Apples, Broccoli, Cherries, Lettuce, Melons, Peaches, Celery, Squash, Cucumbers, Pears, Plums, Beans, Grapes, Peppers, Strawberries, Peas, Tomatoes, Butter, Cheese, Milk, Meat of Cattle, Goat, Hogs and Poultry	Endosulfan (Thiodan)
Butter, Cheese, Milk	Endrin
Blackberries, Blueberries, Cherries, Raisins, Apples, Peppers, Tomatoes, Citrus Fruits, Grapes, Cantaloupes, Wheat, Barley	Ethephon (Ethrel)
Apples, Citrus Fruits, Grapes, Pears, Beans, Peaches, Plums, Strawberries, Tomatoes, Meat of Cattle	Ethion
Poultry, Eggs, Apples, Pears	Ethoxyquin (Santoquin)
Apples, Broccoli, Mushrooms, Grapes, Pears, Lettuce, Peppers, Celery, Tomatoes, Cucumbers	Ethylenebisdithiocarbamate Fungicides, Mancozeb, Maneb, Metiram and Zineb
Apples, Pears, Citrus Fruits, Tomatoes, Cucumbers	Fenbutatin Oxide (Vendex)

Milk	Fenoxapropethyl
Most Berries, Apples, Beans, Broccoli, Carrots, Celery, Cherries, Cucumbers, Grapes, Lettuce, Melons, Peaches, Peanuts, Pears, Peas, Peppers, Plums, Squash, Tomatoes	Ferban
Milk	Flucythrinate
Most Berries, Celery, Apples, Avocados, Cherries, Grapes, Lettuce, Onions, Melons, Tomatoes, Citrus Fruits, Squash, Cucumbers	Folpet (Phaltan)
Citrus Fruits, Apples, Peaches, Pears, Plums	Formetanate Hydrochloride (Carzol)
Soyabeans, Strawberries, Mustard, Milk	Fluazifop-Butyl
Lentils, Rapeseed (Canola), Potatoes, Liver and Kidneys of Cattle, Goats, Hogs, Poultry and Sheep	Glufosinate-Ammonium
Barley, Oats, Soybean, Wheat, Peas, Beans, Lentils, Flax, Kidney and Liver of Cattle, Goats, Hogs, Poultry and Sheep	Glyphosate
Butter, Cheese, Milk, Meat of Cattle, Goats, Hogs, Poultry and Sheep	Heptachlor
Most Nuts/Beans, Barley, Corn, Rice, Rye, Wheat	Hydrogen Cyanide
Grapes, Cherries, Peaches, Tomatoes, Strawberries, Kiwi, Cucumbers, Beans	Iprodione (Rovral)
Citrus Fruits	Imazalil
Apples, Tomato Paste/Purée, Potatoes	Imidacloprid
Milk, Eggs, Rapeseed (Canola)	Isofenphos
Apples, Avocados, Broccoli, Celery, Plums, Cauliflower, Cherries, Cucumbers, Grapes, Lettuce, Melon, Peaches, Pears, Peppers, Pineapple, Spinach, Squash, Strawberries, Tomatoes, Butter, Cheese, Milk, Meat of Cattle, Goats, Hogs, Sheep and Poultry	Lindane
Most Berries, Avocados, Grapes, Melons, Plums, Pecans, Mushrooms, Pineapples, Raisins, Raw Cereals, Asparagus, Cherries, Lettuce, Peaches, Cucumbers, Squash, Tomatoes, Apples, Beans, Pears, Celery, Broccoli, Cauliflower, Peas, Peppers, Beets, Carrots, Garlic, Potatoes	Malathion

Onions, Beets, Carrots, Potatoes	Maleic Hydrazide (MH)
Citrus Fruits, Lettuce, Avocados, Plums, Ginseng, Blueberries, Broccoli, Raisins, Cauliflower, Cherries, Cucumbers, Peas, Peaches, Grapes, Peppers, Soyabeans, Beans, Tomatoes, Carrots, Potatoes, Strawberries, Walnuts, Peanuts, Raspberries, Wheat	Metalaxyl
Broccoli, Lettuce, Peppers, Cauliflower, Cucumbers, Tomatoes, Beans	Methamidophos
Citrus Fruits, Apples, Pears, Cherries, Grapes, Peaches, Plums	Methidathion
Cabbages, Grapes, Lettuce, Citrus Fruits, Apples, Celery	Methomyl (Lannate)
Most Berries, Apples, Apricots, Beans, Asparagus, Beets, Broccoli, Carrots, Cherries, Cucumbers, Grapes, Lettuce, Melons, Mushrooms, Peaches, Peanuts, Pears, Peas, Peppers, Pumpkins, Oats, Spinach, Squash, Tomatoes, Barley, Rice, Rye, Soyabean, Wheat, Meat of Cattle, Hogs and Sheep	Methoxychlor
Potatoes	Metribuzin (Sencor)
Citrus Fruits, Apples, Asparagus, Pears, Broccoli, Cauliflower, Celery, Plums, Lettuce, Peaches, Raspberries, Spinach, Strawberries, Tomatoes	Mevinphos (Phosdrin)
Apples, Pears, Tomatoes	Monocrotophos
Eggs, Apples, Grapes, Wine, Raisins, Milk, Meat of Cattle, Goats, Hogs, Horses, Sheep and Poultry	Myclobutanil
Citrus Fruits, Spinach, Broccoli, Lettuce, Cauliflower, Strawberries, Cucumbers, Beans, Melons, Peas, Peppers, Rice, Tomatoes, Soyabeans	Naled (Dibrom)
Most Citrus Fruits, Apples, Apricots, Blackberries, Beans, Broccoli, Celery, Cherries, Corn, Cucumbers, Lettuce, Mushrooms, Melons, Onions, Pears, Parsley, Peaches, Peppers, Plums, Peas, Spinach, Tomatoes, Squash	Nicotine
Maple Syrup	Paraformaldehyde

Most Berries, Beans, Broccoli, Carrots, Celery, Corn, Cucumbers, Lettuce, Peas, Onions, Peppers, Spinach, Squash, Pears, Apples, Cherries, Citrus Fruits, Grapes, Melons, Peaches, Plums	Parathion
Lettuce, Spinach, Celery, Grapes, Apples, Pears, Peaches, Beans, Cucumbers, Plums, Broccoli, Peppers, Tomatoes, Milk, Meat of Cattle and Poultry	Permethrin
Citrus Fruits, Cherries, Apples, Grapes, Plums, Apricots, Peaches, Pears	Phosalone
Apples, Grapes, Peaches, Pears, Cherries, Blueberries, Plums, Kiwi	Phosmet (Imidan)
Most Berries, Raw Cereals, Almonds, Apples, Beans, Cherries, Grapes, Oranges, Peaches, Peanuts, Pears, Peas, Tomatoes, Dry Codfish	Piperonyl Butoxide
Apples	Pirimicarb (Pirimor)
Hops (Dried), Apricots, Grapes, Peaches, Strawberries, Beans, Citrus Fruits, Plums, Apples, Pears	Propargite (Omite)
Barley, Wheat, Oats	Propiconazole
Most Berries, Raw Cereals, Almonds, Apples, Beans, Cherries, Grapes, Pears, Oranges, Peaches, Peanuts, Tomatoes, Walnuts, Peas	Pyrethrins
Soyabeans, Eggs, Milk, Meat of Cattle, Goats, Hogs, Horses, Poultry and Sheep	Quizalofopethyl (Assure, DPX-Y6202)
Soyabean, Lentils, Beans, Tomatoes, Peas, Onions	Sethoxydim (Poast)
Apples, Pears, Carrots, Peaches, Plums, Citrus Fruits, Cucumbers, Tomatoes, Cherries	Sodium Orthophenyl Phenate (Stop Mould B)
Barley, Oats	Sodium TCA
Milk	Tefluthrin
Apples, Grapes, Meat of Cattle, Hogs and Poultry	Tetrachlorvinphos (Gardona, Rabon)
Cucumbers, Melons, Pumpkins, Plums, Tomatoes, Citrus Fruits, Apples, Pears, Peaches, Grapes, Cherries, Strawberries	Tetradifon (Tedion)

Apples, Citrus Fruits, Pears, Potatoes, Bananas	Thiabendazole
Bananas, Apples, Celery, Peaches, Strawberries, Tomatoes	Thiram
Barley, Wheat	Tralkoxydim
Wheat	Triasulfuron
All Food Crops	Triazolyl Alanine (TA) Plant Metabolite From Triazole Fungicides
Barley, Wheat, Milk	Tribenuronmethyl
Liver and Kidney of Cattle, Goats, Hogs, Horses and Sheep	Triclopyr
Carrots	Trifluralin
Kiwi, Strawberries, Apricots, Grapes, Lettuce, Cherries, Tomatoes, Peaches, Peppers, Beans, Cucumbers, Plums	Vinclozolin
Most Berries, Apples, Apricots, Beans, Beets, Broccoli, Carrots, Celery, Cherries, Cucumbers, Grapes, Lettuce, Melons, Onions, Peaches, Peanuts, Peppers, Peas, Pears, Spinach, Squash, Tomatoes	Ziram

Veterinary Drugs

Authors' Note: *The following list of veterinary drugs shows the large number of drugs, such as antibiotics or growth hormones, which may be administered to animals or added to their feed. The use of these drugs by producers saves time and money in raising animals to market size.*

Scientists are now acknowledging that growth-promoting antibiotics used in indiscriminate ways in animals are leading to the development of dangerous new resistant strains of bacteria. According to an article in the Globe and Mail, *July 13, 1998, a scientific panel made up of health specialists, food industry scientists, disease experts and consumer representatives examined the role of drugs in the food supply and concluded "that animals are developing antibiotic-resistant bacteria and are passing them on to the humans who eat them."*

Our concerns regarding the foregoing were not alleviated when we read the July 11, 1997, Globe and Mail *article which informed us that "The Federal Health Department is closing its in-house drug research laboratories and will rely instead on scientific research from Universities, foreign regulatory agencies and the pharmaceutical companies that apply to have their drugs approved." Furthermore, it appears the downsizing is already in progress. On April 13, 1998, the* Globe and Mail *reported "all food inspection functions formerly done by four government departments have been integrated under the roof of a new organization, The Canadian Food Inspection Agency, that is pulling back from traditional on-site inspection methods and placing more emphasis on self-policing by food processors."*

Milk, Edible Tissue of Swine and Cattle	Ampicillin
Eggs, Muscle, Liver and Kidneys of Chickens and Turkeys	Amprolium
Kidney of Swine	Apramycin
Muscle and Liver of Swine, Turkeys and Chickens, Eggs	Arsanilic Acid
Muscle, Liver, Kidney, Skin and Fat of Chickens	Buquinolate
Milk and Edible Tissue of Cattle	Cephapirin
Kidney, Liver and Muscle of Cattle, Chicken, Sheep, Turkeys and Swine	Chlortetracycline

Muscles, Liver and Kidneys of Chickens and Turkeys	Clopidol
Muscle, Kidney, Liver and Fat of Chickens, Cattle and Goats	Decoquinate
Milk	Dihydrostreptomycin
Muscle, Liver and Fat of Chickens and Turkeys	Dinitolmide (Zoalene)
Milk, Edible Tissue of Chickens, Turkeys and Swine	Erythromycin
Edible Tissue of Turkeys and Kidneys of Swine	Gentamicin
Milk	Hydrocortisone
Liver of Sheep and Cattle	Ivermectin
Edible Tissue of Cattle, Sheep and Swine	Levamisole Hydrochloride
Edible Tissue of Cattle	Monensin
Edible Tissue of Calves	Neomycin
Muscle, Liver, Kidney and Skin of Chickens	Nicarbazin
Muscle and Liver of Turkeys	Nitarsone
Edible Tissue of Cattle, Chickens and Turkeys	Novobiocin
Milk, Edible Tissue of Turkeys and Cattle	Penicillin G
Milk	Polymyxin B
Muscle, Liver and Kidney of Swine	Pyrantel Tartrate
Muscle, Liver, Kidney, Skin and Fat of Chickens	Robenidine Hydro-Chloride
Muscle and Liver of Swine, Chickens and Turkeys, Eggs	Roxarsone
Edible Tissue of Chickens	Spectinomycin
Milk	Streptomycin
Edible Tissue of Cattle and Swine	Sulfachlorpyridazine

Milk and Edible Tissue of Cattle	Sulfadimethoxine
Edible Tissue of Cattle	Sulfaethoxypyridazine
Edible Tissue of Cattle, Swine, Chickens and Turkeys	Sulfamethazine
Edible Tissue of Swine	Sulfathiazole
Edible Tissue of Calves, Swine, Sheep, Chickens and Turkeys	Tetracycline
Milk and Edible Tissue of Cattle, Goats and Sheep	Thiabendazole
Liver of Swine	Tiamulin
Muscle, Liver, Kidney and Fat of Cattle, Swine, Chickens and Turkeys	Tylosin

6

Canada's Labelling Standards for Genetically Engineered Foods

Canada is a member of the Codex Alimentarius Commission, an agency which sets international standards for food production and labelling. The Commission, formed by the World Health Organization (WHO) and the Food and Agriculture Organization (FAO) as a joint venture, is based in Rome. Codex standards have been recognized by the World Trade Organization as the basis for trade agreements and they are becoming quite important for countries who wish to conduct international trade. In 1995, Codex approved a plan which called for the establishment of guidelines for labelling of foods derived from biotechnology. The Codex Committee on Food Labelling (CCFL) was appointed to accomplish this task and their mandate was to establish legally binding international labelling standards.

Since 1993, there have been three major Codex consultations on novel foods derived from genetic engineering. Canada's major trading partners, including the United States and Japan, support labelling on a case by case basis only when there are concerns about the health, safety and compositional change of the food. Health Canada, who is responsible for setting our food labelling policies, felt our national guidelines for labelling novel foods should be done in conjunction with standards developed at the international level.

A proposed definition of novel foods was published by the Health Protection Branch, Food Division, in the *Canada Gazette Part I*, on August 26, 1995, which is the first step on the road to becoming law. The most pertinent and worrisome sections of the novel foods definition are clearly laid out here. A "novel food" is

(a) a substance, including a microorganism, that has not previously been manufactured for use as food in Canada,

(b) a food, including a food additive, that is manufactured by a process that has not previously been used to manufacture that food, where as a result of that process

(i) the food exhibits characteristics that were not previously observed in that food, or

(c) a food that is derived from a plant, animal or microorganism that has been genetically modified such that

(i) the plant, animal or microorganism exhibits characteristics that were not previously observed in that plant, animal or microorganism.

As a result of the Codex consultations, Canada developed a set of guidelines that reflected a general international agreement to

- require mandatory labelling if there is a health or safety concern, [for example] from allergens or a significant nutrient or compositional change.
- ensure labelling is understandable, truthful and not misleading.
- permit voluntary positive labelling on the condition that the claim is not misleading or deceptive and the claim itself is factual.
- permit voluntary negative labelling on the condition that the claim is not misleading or deceptive and the claim itself is factual.

These principles are consistent with policy for all foods under the *Food and Drugs Act*.

These guidelines are set out in a bulletin from the Office of Biotechnology. The bulletin goes on to explain the difficulties of mandatory labelling of genetically engineered products. *See also* B.01.056., a new regulation added in the last year that addresses the subject of novel foods.

While mandatory labelling will be required when the genetically engineered products present a health or safety concern, the government recognizes the consumers' desire for more non-safety related information. There are practical difficulties and costs associated with tracking foods in the food system to know whether or not they come from a genetically engineered crop, or contain any ingredients that may have been genetically engineered. This raises serious concerns about enforcement. How could governments offer assurance that labels were correct? In most cases there is no way of distinguishing a genetically engineered product from another, so unless the genetically engineered product has been carefully tracked from the farmer's field, they would be unidentifiable to a government inspector.

There is also a concern that so many foods would have some ingredient or component that has been genetically engineered that the labels would become meaningless. Most processed foods would soon have to be labelled as "may

> contain" some product of genetic engineering. There is also a concern that this kind of labelling on so many products might take away from critical health information on a label.

This bulletin is an insult to any consumer's intelligence. Ottawa did not specify the exact nature of the critical health information and we assume they were referring to calorie content or nutritional content. If that is the case, we can assure them we would rather a label declare whether the food has been genetically engineered in lieu of calorie content or nutritional content.

The biotech industry promotes the belief that genetic engineering and normal plant breeding are one and the same and that foods made in this way are identical in essential makeup to normal crops or are "substantially equivalent." (Accordingly, government officials claim they will encounter difficulty in tracking or distinguishing genetically engineered products. We recommend they contact Genetic ID, a company who claims they can "positively detect genetically modified content in crops and foods." They are located at Suite 208 – 500 North Third Street, Fairfield, Iowa, 52556. Phone: 515–472–9979 Fax: 515–472–9198.) The Plant Biotechnology Office defines substantial equivalence as

> the equivalence of a novel trait within a particular plant species, in terms of its specific use and safety to the environment and human health, to those in that same species that are in use and generally considered as safe in Canada, based on valid scientific rationale.

In order to better understand this definition, let's consider what this means for potato plants, for example. One plant is genetically engineered and the other is not. They are basically the same plant, they look the same, produce potatoes that are supposed to taste the same and are used in the same way. Therefore, Health Canada and the biotech industry claim they are substantially equivalent. This means these genetically engineered potatoes do not have to be kept separate from the non-genetically engineered potatoes during shipping or storage, nor do they have to be identified as such on their labels. Our concern is that should the genetically engineered potato contain an uncommon allergen or a toxic by-product as was the case with tryptophan (*see* page 114), it would make it extremely difficult to trace the toxin back to the potato. Furthermore, in the absence of mandatory labelling laws, if a company turns their genetically engineered potatoes into a product such as french fries, the consumer would have no way of knowing that the french fries were made from genetically engineered potatoes. Health Canada's choice to follow international standards rather than writing "Made in Canada" food laws has thereby removed our democratic right to freedom of choice.

The following is a recommendation from the Codex Committee meeting held in Ottawa in 1997. Our concern with this recommendation is that these genetically modified organisms (GMOs) are already in our food supply and that Codex is only now suggesting that experts be consulted to determine their dangers, rather like putting the cart before the horse.

> The Consultation considered gene transfer from GMOs, and as likelihood of transfer **from a genetically modified plant to a micro-organism in the gastro-intestinal tract is remote but cannot be entirely ruled out**, the Consultation recommended that FAO/WHO convene an expert consultation to address whether there are conditions or circumstances in which antibiotic-resistance marker gene(s) should not be used in genetically-modified plants intended for commercial use.

This next shocking statement is taken from the same report of the Codex Alimentarius Commission. Here they are discussing the consumers' right to know whether foods have been produced by biotechnology. Read this and ask yourself how can one possibly "ill-define" the consumers' right to know? Consumers want to know which foods are genetically engineered and want them labelled as such. Is it possible for consumers to be any clearer than that? The report states:

> [The Codex Alimentarius Commission] **noted** the opinion **that, while consumers may claim the right to know whether foods had been produced by biotechnology, this right was ill-defined** and variable **and in this respect could not be used by Codex as the** primary **basis of decision-making on appropriate labelling. . . . Foremost among these was the protection of consumers' health from any risks introduced** by the production process, **followed by nutritional implications resulting from changes** to the composition of the food, any significant technological changes in the properties of the food itself, **and the prevention of deceptive trade practices**. To a considerable extent **such matters would have to be decided on a case-by-case basis. The Executive Committee noted that the possibility of voluntary labelling always existed.**

The latest meeting of the CCFL was held in Ottawa in May 1998. Representatives from the European Union, the majority of Asian countries as well as consumer groups from more than one hundred countries attended the meeting. They were urging delegates to adopt the proposal that all genetically modified foodstuffs be labelled as such. Unfortunately, representatives of the food industry dominated the conference and favoured labelling of genetically modified foods only if there is a proven health risk or a change in the nutritional content of the food. Canada and the US back this position as multinational corporations in Canada and the US are major producers of genetically engineered soyabean and canola. This meeting was an unsuccessful attempt by the Codex Committee to find a compromise satisfactory to opposing camps.

We conclude from this that the Codex Committee and the biotechnology industry are seeking only the appearance of a clear and stringent labelling system while avoiding labelling genetically engineered foods.

Dr. John Fagan, a biology expert, commented, "At present, the Executive Committee of Codex has proposed a policy on labelling that is highly favourable to the biotechnology industry, but fails to protect consumers."

Consumers International's Director General, Julian Edwards, whose delegates attended the meeting said, "After years of discussion in Codex, consumers still don't have the information they need and meanwhile, month by month, new genetically engineered foods are being introduced into the marketplace."

Canada's laws will continue to be mediocre half-measures until concerned consumers speak out. If you would like the government to legislate mandatory labelling of all genetically engineered products, please fill out and submit the petition at the back of this book. It is not too late!

7

Field Testing Genetically Engineered Seeds and Plants

The following pages will give you a glimpse of the way manufacturers are required to field test genetically engineered seeds and plants and shows some of the restrictions under which they must conduct their testing. As you will see, some of the questions manufacturers are required to answer are very macabre.

This section provides you with portions of *Regulatory Directive 95–01, the Guidelines for Submission of Applications For Authorization of Confined Field Trials of Plants With Novel Traits (PNTs)*. It is provided by Ottawa to help manufacturers prepare their applications and keep processing delays to a minimum. As these plants and seeds are to be tested for the first time in an agricultural setting, Ottawa feels it must identify areas of potential concern and the information supplied in these applications will help Health Canada assess the environmental effects of unconfined release of plants with novel traits. It must, therefore, include details of other life forms with which the PNTs will interact. Health Canada willingly acknowledges these assessments will be part of a continuum of research, development, evaluation and commercialization of plants with novel traits. It is admittedly a "learn as you go" process which does not serve the best interests of humanity or the environment.

Agriculture and Agri-Food Canada has been regulating the field testing of plants with novel traits since 1988. Since then, the government has approved over 2,000 field releases of genetically engineered crops. This number is much greater than we ever anticipated. The following list shows some of the genetically engineered products that have already been through the field testing process and which have received approval to be sold as foods in Canada. Note the companies that appear over and over again.

Product	Manufacturer	Date
Imidazolinone-resistant maize	Pioneer Hi-Bred Intl.	May 1994
Glyphosate-tolerant-canola	Monsanto Canada	November 1994
Imidazolinone-resistant-canola	Pioneer Hi-Bred Intl.	April 1995
Colorado potato beetle-resistant potato	Monsanto Canada	September 1995
Delayed ripening tomato	DNA Plant Technology	November 1995
Insect-resistant corn	CIBA Seeds	December 1995
Glyphosate-tolerant-soyabeans	Monsanto Canada	April 1996
Insect-resistant cotton	Monsanto Canada	April 1996
Canola	Calgene	April 1996
Processing tomatoes - exhibiting reduced pectin degeneration	Zeneca Plant Science	June 1996
High oleic acid/Low linolenic acid canola	Pioneer Hi-Bred Intl.	August 1996
Cotton	Calgene	August 1996
Insect-resistant cotton	Monsanto Canada	Novemeber 1996
Colorado potato beetle-resistant potato	Monsanto Canada	Novemeber 1996
Glufosinate-tolerant corn	DEKALB Genetics Corp.	December 1996
Glyphosate-tolerant cotton	Monsanto Canada	December 1996
Insect-resistant corn	Pioneer Hi-Bred Intl.	December 1996
Glufosinate ammonium-tolerant canola	AgrEvo Canada	February 1997
Insect-resistant corn	Monsanto Canada	February 1997
Glufosinate ammonium-tolerant corn	AgrEvo Canada	April 1997
Sethoxydim-tolerant corn	BASF Canada	February 1997
Insect-resistant- and glufosinate-tolerant corn	DEKALB Genetics Corp.	April 1997
Imazethapyr-tolerant corn	Zeneca Seeds	July 1997
Bromoxynil-tolerant canola	Rhône-Poulenc Agriculture	July 1997
Glyphosate-tolerant canola	Monsanto Canada	September 1997
Insect-resistant and glyphosate-tolerant corn	Monsanto Canada	September 1997
Virus-resistant squash	Seminis Vegetable Seeds, Inc.	April 1998
Imazolinone-tolerant corn	Pioneer Hi-Bred Intl	June 1998

Here is the process each of these companies must go through each time they want to introduce a plant with a novel trait.

Guidelines For the Submission of Applications for Authorization of Confined Field Trials of Plants With Novel Traits (PNTs)

A2.1 **The scope of this document covers all plants with novel traits** that are of concern (i.e., **that are not familiar or substantially equivalent to products already on the market**, in use and generally regarded as safe).

A2.4 **Due to the broad range of PNTs that are being developed, this document** and its guidelines **should be considered flexible and will likely evolve as more experience is gained.** A determination of the need for unconfined release authorizations and the environmental safety assessment of PNTs will be conducted on a case-by-case basis, founded on familiarity and substantial equivalence.

A4.0 Applications that are complete will allow for timely evaluations with a minimum of correspondence requesting further information.

B3. **A covering letter must be sent with the application, summarizing the intent of the trials**, and including the plant species, the modified trait(s), the number of trials applied for, the general trial locations, **and whether non-registered pesticides will be used, or registered pesticides will be used for non-registered purposes.**

B3.2 Please indicate in the application **which of the following information is to remain confidential business information:**

- exact trial site
- plasmid map(s)
- exact genetic change

B4.0 Please indicate in the application **which information is to remain confidential business information.** However, please **do not apply a confidential stamp to cover all pages of the application, unless you really consider all included information on a page to be confidential.**

In completing the application, **applicants may consider it unnecessary or inappropriate to provide certain information.** In these instances, **information requirements may be waived if valid scientific rationale is provided.**

C3.6 **The toxicity of the novel gene products**, breakdown products **and by-products in the environment must be established.** Describe:

1. **potential toxigenicity to known or potential predators, grazers, parasites, pathogens**, competitors and symbiont;
2. **potential for adverse human health effects, e.g., exposure to toxins, irritants and antigens. Include estimated level and most likely route of human exposure to the gene products, breakdown products and by-products.**
3. Biochemistry: **for novel gene products identified in Part C3.6 that are known to be toxic, describe:**
   a. **likelihood and change of level of exposure of consumers** and symbiont;
   b. **the effect on soil micro flora and fauna.** Residual studies may be conducted to determine macro changes. Observed changes at this level may warrant further in-depth studies.

C4.1 **What trait(s) are conferred to the recipient plant?**

C4.2 **Was the modification achieved through mutagenesis, recombinant DNA transformation,** or other means?

C5.4 **Once inserted into the plant, has each genetic modification been shown to be stable?** Please submit data demonstrating stability.

C5.6 **Is the modified material known to produce toxins that were not produced by the unmodified material?**

If a researcher answers "yes" to this question, will the gene altered plant still gain approval? Specific independent testing should be required to be certain that toxins are not produced.

C5.7 **Provide data showing the fate of the gene products when ingested by:**
a. **Humans or livestock**
b. Native faunal populations (i.e., mammals, birds, reptiles and insects)

C5.8 If no data are available, describe the probable fate of the gene products when ingested by:
a. Humans or livestock
b. Native faunal populations (i.e., mammals, birds, reptiles and insects)

C5.7 and C5.8 above appear to be saying "If you have the data, show it to us. If you don't have the data, just fake it." It does not seem particularly scientific to "describe the probable fate." The data referred to in C5.7 should be mandatory.

C6.1 **Supply a map of the trial site showing general geographic location,** specific legal description, exact location of trial plots, the experimental trial number **and, whenever possible, crops that are grown within the isolation distance.**

C7.3 **What reproductive isolation measures are proposed?** Describe fully, for each site:
a. Isolation distances
b. The use, size and arrangement of border/guard rows.
c. **The use of any physical methods to prevent pollen movement** (e.g., cages, bags or flower removal).

C7.4 Seeding:
a. How much seed is to be planted?
b. How will the trial material be packaged and transported to the trial site?
c. Will the material be planted by hand or machine?
d. **What precautions will be taken to avoid dissemination of seed from the trial site?**
e. Is it proposed to plant any unmodified plants of the same species, or related species, to determine, for instance, herbicide efficacy?
f. What will be the fate of excess seed from planting?
g. What procedures are in place for recording quantities of surplus seed?

C7.5 Spraying:
a. Will the trial be sprayed with pesticides?
b. If so, are the pesticides registered products?
c. Are they registered for use on the trial material?
d. What total surface will be sprayed?

C7.6 Harvesting:
a. **Will the plants be allowed to set seed?**
b. If so, will the seed or other harvested product be harvested by hand or machine?
c. **If by machine, what precautions will be taken to avoid dissemination of seed from the trial site?**
d. **How will remaining plant matter be disposed of?**

e. What will be the fate of retained seed (if any)? How will it be stored if retained?

f. What procedures are proposed for recording quantities of harvested material?

C7.9 What contingency plans have been developed in the case of accidental release of seed or propagules?

C7.10 What contingency plans have been developed in the case of unexpected spread of the PNT material?

C8.1 What monitoring procedures are proposed during the trial period?

C8.2 What monitoring procedures are proposed during the post-trial period?

C9.1 Has there been any public notice of the proposed field trial(s)?

C9.2 Has there been any concern expressed by the public, or other persons, concerning the proposed field trial(s)?

C9.3 Is public notice and or media attention proposed once the trial is approved? Public notice of confined trials is recommended. The means by which this notice occurs is left to the applicant. We recommend announcements to immediate neighbours and press releases to local newspapers.

Notification to the public should be made mandatory, not just recommended. The notification should also be "plant name specific" so that people know exactly which plants are being planted in proximity to other crops. They should also be told what steps are being taken to contain the gene-flow. Furthermore, if a farmer is absolutely opposed to these plants being grown next to his farm because of his concern for gene flow, then that should be sufficient reason for the company to find another site.

D1.2 Agricultural-Silvicultural Practices:

1. What are the proposed release sites for the PNT?
   a. All of Canada?
   b. Specific regions?
   c. What is the projected area (hectares) of release?

Where there is potential for gene flow from the PNT into related species, detail the consequences of novel gene introgression into those species and resulting expression.

Post-release monitoring must be carried out by the applicant. If, at any time after authorization for unconfined release, the applicant becomes aware of any new information regarding risks to the environment (e.g., enhanced weediness characteristics) or human health (e.g., exposure to allergens) that could be caused as a result of the release, the applicant must immediately inform the Director of the Plant Products Division.

The following recommended requirements are designed to minimize the transfer of novel genes, through pollen transfer, from PNTs to plants of related crops or wild species. Major crops in Canada, such as canola, may hybridize with related wild species. Some of the traits that are introduced into these crops are traits that might enhance the "weedy" characteristics of these weeds.

A number of precautions can be employed to mitigate the possibility of gene transfer and weediness. The required isolation distances correspond to minimal requirements for a given species. More stringent conditions may be imposed depending on the novel trait.

Health Canada does not appear to have considered what would happen should nature provide us with a flood, windstorm or tornado – isolation distances would then be irrelevant.

For the purpose of this document, weeds are considered a subset of plants that may be considered pests. The term "weed" is used to describe a plant species that is a nuisance to humankind in that it occurs in "managed" ecosystems where it is unwanted. Weeds tend to spread easily in disturbed areas or among crops. Whereas it can be considered that any "plant out of place" is a weed, the emphasis of these guidelines is to determine whether a PNT could be successful in colonizing managed ecosystems at the expense of displacing other species, in particular cultivated crop plants.

D4 Other Requirements

- **In order to prevent dissemination of seeds of PNTs, all seeding and harvesting equipment must be cleaned on site before removal to another location.**
- For some species, plants must be harvested before full maturity to prevent seed dispersal.
- An acceptable monitoring protocol must be established.
- **A protocol must be established to recover seeds in case of accidental release.**
- **Leftover and progeny seed, including residue removed from equipment**

cleaning and seeds from guard rows, must be destroyed by heating, burning, or deep burial.
- No seed or its progeny, including those from guard rows, must enter human or livestock food/feed chains without a food safety assessment by Health Canada.
- **Plant residues, remaining after the trials are completed, must be soil incorporated.**
- **A detailed seed and plant log must be kept, available for inspection by AAFC staff, to track distribution and disposition of seed and progeny of each trial.**
- **A report of the conduct of the trial must be submitted to AAFC.**

Does Agriculture Canada really believe that every corporation will follow all of these regulations to the letter? To us, they seem far-fetched, especially when one considers that each company producing genetically engineered seeds and plants may have many field trials taking place simultaneously. Imagine hectares of fields planted with crops containing novel genes. Then picture each plant with a little bag or cage over its head to prevent gene flow as is being recommended in C7.3. This takes a vast stretch of the imagination to believe and we doubt whether a company would ever go to this extent.

Although the following is taken from a separate directive, *Regulatory Directive 94-08, Assessment Criteria for Determining Environmental Safety of Plants With Novel Traits*, we felt it important enough to include here. Health Canada requires companies to explain the methods used to

- introduce novel traits
- list, identify the source and describe genes, including antibiotic resistance, other marker genes or regulatory genes.

They also require the following information about vectors:

If a vector was employed:

i. what is the vector name and cloning method?
ii. is the vector naturally pathogenic?
iii. was the vector disarmed?
iv. how was the vector disarmed?
v. is there expression of the gene in the vector?

We don't believe that scientists can guarantee a naturally pathogenic vector is completely disarmed. If Ottawa must ask these questions of researchers, the potential for harm must exist. The consumer is expected to take on faith that researchers never make mistakes and to accept that the organisms they are transferring into food products have been judged completely harmless based on something as vague as "valid scientific rationale."

Health Canada should not be making decisions based on research supplied by the companies who are producing the product. There is a great deal of room for conflict of interest here. Any self-serving company could falsify data in order to market their products more expediently.

The following Agriculture Canada document shows how, Monsanto's Roundup Ready™ seed, was evaluated using the aforementioned Regulatory Directives. In this way Monsanto was able to receive government approval for the use of Roundup Ready™ (seeds which will greatly increase the use of glyphosate, a dangerous herbicide). We find the actions of Agriculture Canada bizarre given the significant scientific research available that proves the dangers of this chemical. See Genetically Engineered Foods page 118 for further information on Roundup-Ready™ seeds and glyphosate.

Decision Document DD95-02: Determination of Environmental Safety of Monsanto Canada Inc.'s Roundup™ Herbicide-Tolerant Brassica napus *Canola Line GT73*

Agriculture and Agri-Foods Canada (AAFC), specifically the Plant Biotechnology Office and the Feed Section of the Plant Products Division, **have evaluated information submitted by Monsanto Canada Inc.** regarding the canola line GT73. This line has Roundup-Ready genes that express novel tolerance to glyphosate, the active ingredient of Roundup herbicide. **The Department has determined that this plant does not present altered environmental interactions** when compared to existing commercialized canola varieties in Canada, and **is considered substantially equivalent to canola** currently approved as livestock feed.

Unconfined release into the environment, including feed use of GT73, and other B.napus lines derived from it, but without the introduction of any other novel trait, **is therefore considered safe.**

According to extensive data initially submitted to Health Canada, using simulated digestive fluids and acute mouse gavage studies, **the enzymes** expressed by the Roundup-Ready™ genes **are rapidly inactivated and degraded in the gastric and intestinal systems.**

Two genes have been introduced into the variety "Westar," which in combination provide field level tolerance to glyphosate, the active ingredient in Roundup herbicide. **The exact nature of these genes is considered confidential business information by Monsanto.**

You can request a copy of the bulletin *Field Testing Plants With Novel Traits in Canada 1995* from Plant Products Division, Food Production and Inspection Branch, 59 Camelot Drive, Nepean, ON, K1A 0Y9.

There is growing resistance worldwide to field testing genetically engineered plants – and with good reason. These directives state that it is up to the companies planting these crops to identify their effects on animals and humans and to take appropriate steps to notify the proper government department. Joy Bergelson, professor of Ecology and Evolution at the University of Chicago was quoted in *The Globe and Mail* as saying: "Findings show that genetic engineering can substantially increase the chances of transgene escape or the spread of certain traits from one plant to another." Her findings showed a dramatically heightened ability in gene-altered plants to reproduce sexually and spread their modified genes to traditional plants. In fact, in some cases the genetically altered plants fertilized other plants twenty times faster than normal. Researchers are not sure why gene-altered plants become more fertile but Bergelson speculated that "the pollen from the genetically altered plants might have a longer life span than normal pollen or have some other competitive advantage."

In recent years, Health Canada has been re-evaluating its role and obligations to Canadians as protector of public health. This role has been seriously affected by cutbacks. The 1993/94 Health Protection Branch budget was $237 million. The projected budget for 1999/2000 is half of that, $118 million. To maintain budget projections, Health Canada has divested itself of direct involvement in many sectors related to health and food. Health Canada now sees itself as an overseer with little direct involvement – and this is very cost effective! By dismantling food and drug laboratories and contracting out research and testing services, critics claim that Ottawa has completely removed itself from any future legal liability in these areas. The excerpts from the *Food and Drugs Act* included in this book will certainly attest to that. It is essential for Canadians to raise their voices NOW and demand a moratorium on the release of these plants and seeds until all the implications are fully understood.

Glossary

Acidulant – a substance used to make foods more acidic. May also be used as a preservative.
Additive – *See* Food additive.
Aerating agent – *See* Propellant.
Alkali – a substance used to make foods less acidic or more alkaline.
Anticaking agent – a substance used to prevent certain types of food from becoming lumpy.
Anti-foaming agent (defoamer) – a substance used to prevent excessive foaming or film build-up in products such as jam.
Antimicrobial agent – an additive that prevents the growth of bacteria, moulds and yeast in foods.
Antioxidant – a chemical additive that prevents food from reacting with oxygen in order to maintain its odour, taste and colour.
Astringent – a substance used medically to cleanse the skin and to constrict the pores.
Biodiversity – the presence of a variety of life forms, from one-celled fungi, protozoa and bacteria to complex organisms such as plants, insects, fishes and mammals, in an environment.
Bleaching agent (maturing agent) – a chemical additive used to whiten flour. Naturally matured flour takes a longer time to lose its gold colour and turn white than flour treated with a bleaching agent.
Breakdown product – a product or part of a chemical which is left after it has been broken down or metabolized.
Cancer-inciter – a substance that on its own may not cause cancer but may have toxic effects on cells which contribute to the development of cancer.
Carcinogen – a cancer-causing agent.
Carrier – *See* Extraction solvent.
Catalyst – a substance that initiates or quickens the rate of a chemical reaction.
Cathartic – a substance used to induce vomiting and/or diarrhea.
Chelator – a substance used to prevent food from discolouring or going rancid by trapping trace metals.
Coating agent – a chemical additive used to coat the skins of certain fruits and vegetables to minimize bruising during transportation and to prevent mould and dehydration.
Co-carcinogen – a substance that does not cause cancer on its own but increases the body's sensitivity to specific carcinogens.
Colour lake dispersion – is formed by mixing dyes with alumina hydrate which keeps the dyes insoluble.
Confined Field Testing – conditions of confinement, for growing plants with novel traits, which include isolating the plants, monitoring the site and restricting land use after the harvest.
Defoamer – *See* Anti-foaming agent.
Delaney Amendment – an amendment to a 1958 law which stipulated that US food chemical manufacturers had to test additives and submit the results to the FDA before the additives were put on the market. The amendment specifically states that no additive may be permitted in any amount in food if it causes cancer when consumed by humans or animals. Ever since it was enacted, the food and chemical industries have tried to get it repealed. By the time you read this, they may have already succeeded. The amendment was written by Congressman James Delaney.
Diuretic – a substance that causes increased output of urine.
Dough-conditioning agent – a substance used to make dough easier to handle and to improve the texture and volume of bakery products.
Dusting agent – *See* Releasing agent.
Emulsifying agent – a substance that facilitates the combination of two or more liquids which would normally separate, such as water and oil.

Enzyme – any one of a large class of proteins produced by living cells. Enzymes act as catalysts to produce specific chemical changes in other substances.
Extraction (carrier) solvent – a substance used to isolate or extract components such as colours and flavours from certain foods to be used in other foods. The decaffeination process can also make use of these solvents to extract caffeine from coffee beans.
FAO/WHO Expert Committee on Food Additives – FAO stands for Food and Agricultural Organization and WHO stands for World Health Organization. Together they form an international committee with members from the government, industry and the university scientific community, which evaluates the safety of certain food additives and then publishes its findings and recommendations. FAO/WHO doesn't perform any tests. They rely on results of scientific data supplied to them.
FDA – *See* Food and Drug Administration.
Field trial – growing and testing a plant species in an agricultural setting.
Filler – a substance added to certain processed foods to increase their bulk or weight or to dilute expensive ingredients.
Firming agent – a chemical additive used to control the desired firmness and texture in foods.
Food additive – any chemical, natural or synthetic, that is added during the manufacture of food.
Food and Drug Administration (FDA) – a branch of the United States Department of Health, Education and Welfare founded in 1906 to safeguard consumers from dangerous foods, drugs and cosmetics.
Gelling agent – a substance used to produce a gel in products such as gelatin desserts.
Generally Recognized As Safe (GRAS) – a US food additive category established in 1958 to exempt additives which were considered at the time to be unquestionably safe, in order to avoid costly toxicological testing as required by law. In recent years, developments in science and in consumer awareness have brought to light the inadequacies of the FDA testing of food additives and, ironically, the complete lack of testing of the GRAS category. There are numerous additives that have toxic effects yet are still on the GRAS list.
Genetically modified organism (GMO) – an organism that has had its DNA altered.
Genotoxic – toxic to the genes
Genotype – the genetic constitution of an organism.
Glazing agent – a chemical additive used to give certain foods a shiny surface.
GMO – *See* Genetically modified organism.
GMP – *See* Good Manufacturing Practice.
Good Manufacturing Practice (GMP) – a regulation in the Canadian *Food and Drugs Act* that allows manufacturers to use as much of an additive as they feel is necessary to achieve a desired effect.
GRAS – *See* Generally Recognized As Safe.
Hepatotoxic – toxic to liver cells.
Humectant – an agent that absorbs moisture or helps another substance retain moisture.
IARC – *See* International Agency for Research on Cancer.
International Agency of Research on Cancer (IARC) – a United Nations organization that gathers information on suspected environmental carcinogens and summarizes available data with appropriate references.
Maturing agents – *See* Bleaching agents.
Metabolism – all the processes by which food taken into the body is transformed into energy.
Modifying agent – a substance used to modify or change the properties of starch.
Mutagen – an agent that induces genetic mutation.
Mutagenic – possessing the ability to create mutations.
Naming the flavour – a term used in the Canadian *Food and Drugs Act* which requires a manufacturer to name the specific flavour used in certain products, for example: (chocolate) milk.

Neutralizing agent – a chemical used to adjust the acidity or alkalinity of certain foods to make processing easier and to ensure consistent results.
Oxidizing agent – a substance used to combine oxygen with another substance.
Parts per million (ppm) – a measure of the amount of a substance in a food or chemical. The Canadian *Food and Drugs Act* states: "parts per million means parts per million by weight unless otherwise stated."
Pathogen – an organism capable of causing disease in another organism. Some common pathogens are bacteria and viruses.
pH – a measure of the relative acidity or alkalinity of a substance.
pH adjusting agent – a chemical additive used to control the amount acidity in a food which could effect the flavour, texture, etc.
PNT – *See* Plant with Novel Traits.
Plant with Novel Traits (PNT) – a plant that has been altered by a specific genetic change and now exhibits characteristics that are different from other plants of its type.
Preservatives (Class I, II, III, IV) – chemicals which are added to foods to prevent or delay spoilage due to the growth of mould, yeast, bacteria and oxidation.
Processing aid – a substance that is added to a food to ensure consistent results and to facilitate processing.
Propellant – a compressed gas used in products such as whipping cream to expel the contents from the containers.
Recipient cell – a cell that receives a new gene.
Releasing agent (Dusting agent) – a substance that prevents sticking. In the case of shellfish, a releasing agent would ease the removal of the meat from the shell.
Sequestering agent (sequestrant) – a substance that combines with and neutralizes trace amounts of metals such as iron and copper. Their use in foods such as salad dressings and alcoholic beverages helps to prevent changes in colour, flavour and texture caused by these metals.
Sequestrant – *See* Sequestering agent.
Stabilizing agent – a substance used to prevent particles in a product from settling out.
Standardized food – a food for which Health Canada has enforced a standard indicating the ingredients the food is allowed to contain and limits for the quantity of each ingredient.
Subacute – the stage in the progress of a disease in which it is not readily apparent.
Substantial equivalence – a term Health Canada uses to denote that a plant with novel traits is equivalent to others in the same species in terms of its use or safety.
Teratogen – a substance that causes birth defects.
Toxic – poisonous.
Transgene – a gene which is taken from one species and inserted into a member of another species.
Transgenic organism – an organism which has had genetic material from a sexually incompatible organism inserted into it.
Unconfined release – growing plants with novel traits in a setting that does not include isolating the plants from surrounding habitats, monitoring the site and/or restricting land use after the harvest.
Unstandardized food – a food for which Health Canada has not set a standard list of ingredients or limits for the quantity of each ingredient a manufacturer may use.
Vector – DNA from an organism into which a gene can be incorporated, that is then introduced into a cell in a foreign organism.
Wetting agent – a chemical additive that enables liquids to be absorbed or to be spread more easily on surfaces.
Wound debriding agent – a substance that dissolves dead tissue in order to prepare the wound for healing.

Bibliography

Agriculture and Agri-Food Canada, Plant Biotechnology Office Variety Section, Plant Products Division, "Assessment Criteria for Determining Environmental Safety of Plants With Novel Traits, Regulatory Directive Dir 94–08," December 16, 1994.

———. "Field Testing Plants With Novel Traits in Canada", 1995.

Associated Press and Reuters. "'Superweeds' Resist Pesticide," *The Globe and Mail*, September 3, 1998.

Balch, James F. and Phyllis A. Balch. *Prescription For Nutritional Healing*. Garden City Park, New York: Avery Publishing Group, 1997.

Barnett, Donna. "Dr. Alexander Morrison, At Your Civil Service," *Harrowsmith*. No. 16, Vol. III.

Blaylock, Russell L. *Excitotoxins: The Taste That Kills*. Santa Fe, NM: Health Press, 1994.

Cab International. *Natural Antimicrobial Systems and Food Preservation*. Guildford, UK: Biddles Ltd., 1994.

Campbell, Murray, "Fears About Food Safety on the Rise," *The Globe and Mail*, April 13, 1998.

Carson, Rachel. *Silent Spring*. New York, NY: Fawcett World Library, 1970.

Cheraskin, E., W. M. Ringsdorf, Jr. and Arline Brecher. *Psychodietetics: Food as the Key to Emotional Health*. New York, NY: Stein and Day, 1974.

Cleave, T. L. *The Saccharine Disease*. Bristol, England: Keats Publishing, Inc., 1975.

Code of Federal Regulations. Parts 1–199. US Government Printing Office, Washington, 1978.

Eggerson, Laura, "Federal Labs to be Shut Down," *The Globe and Mail*, July 11, 1997.

Epstein, Samuel S. *Politics of Cancer*. San Francisco: Sierra Club Books, 1978.

Feingold, Ben F. *Why Your Child is Hyperactive*. New York, NY: Random House, 1975.

Fuller, John Grant. *200,000,000 Guinea Pigs, New Dangers in Everyday Foods, Drugs and Cosmetics*. New York, NY: Putnam, 1972.

Gottschall, Elaine. *Breaking The Vicious Cycle: Intestinal Health Through Diet.* Baltimore, ON: Kirkton Press Ltd., 1994.

Griggs, Barbara. *The Food Factor*. Markham, ON: Penguin Books Canada Limited, 1986.

Hall, Ross Hume, "What Kind of Food Protection System Have We?" *Entrophy Institute Review*, January/February, 1979.

———. *Food for Nought: The Decline in Nutrition*. Baltimore, MD: Harper and Row Publishers Inc., 1974.

Harte, John, Cheryl Holdren, Richard Schneider and Christine Shirley. *Toxics A to Z*. CA: University of California Press, 1991.

Health Canada. *Food Additive Pocket Dictionary*. Minister of Supply and Services Canada, 1996.

———. *Food and Drugs Act and Regulations.* Minister of Supply and Services Canada. 1997.

Ho, Mae-Wan. *Genetic Engineering: Dream or Nightmare?* Bath, UK: Gateway Books, 1998.

Holmes, Randee. *What Have They Done to Our Food?* Toronto, Ontario: McClelland and Stewart, Inc., 1994.

Hubbard, Ruth and Elijah Wald. *Exploding The Gene Myth*. Boston, MA: Beacon Press, 1993.

Hunter, Beatrice Trum. *Consumer Beware! Your Food and What's Been Done To It*. New York, NY: Simon and Schuster, 1971.

———. *Fact Book on Food Additives and Your Health*. New Canaan, CT: Keats Publishing Inc., 1972.

———. *Food Additives & Federal Policy: The Mirage of Safety*. New York, NY: Scribner, 1975.

———. *The Great Nutrition Robbery*. New York, NY: Charles Scribner's Sons, 1978.

Jacobson, Michael F. *The Complete Eater's Digest and Nutrition Scoreboard.* Garden City, NY: Anchor Press/Doubleday, 1985.

———. *Eater's Digest: The Consumer's Factbook of Food Additives*. New York: Doubleday, 1972.

Jones, Mark M., John T. Netterville, David O. Johnston and James L. Wood. *Chemistry: A Brief Introduction*. Philadelphia, PA: W. B. Saunders Co., 1969.

Kneen, Brewster. *From Land To Mouth: Understanding The Food System.* Toronto, ON: NC Press Ltd., 1993.

———. *The Rape of Canola.* Toronto, ON: NC Press Limited, 1992.

La Leche League International. *The Womanly Art of Breastfeeding.* New York, NY: Plume, 1997.

Marine, Gene and Judith Van Allen. *Food Pollution: The Violation of Our Inner Ecology.* First Edition. New York, NY: Holt, Rinehart and Winston, 1972.

Mather, Robin. *A Garden of Unearthly Delights: Bioengineering and the Future of Food.* New York, NY: Penguin Books, 1995.

The Merck Index. Ninth Edition, Rahway, NJ: Merck and Co., Inc., 1976.

Office of Biotechnology, Canadian Food Inspection Agency, *Information Bulletin*, May, 1998.

Reuben, David. *Everything You Always Wanted to Know About Nutrition.* New York, NY: The Maleri Corp., Simon and Shuster, 1978.

Roberts, H.J. *Aspartame (Nutrasweet) Is it Safe?* Philadelphia, PA: The Charles Press, 1990.

Salaman, Maureen and James F. Scheer. *Foods That Heal.* CA: Statford Publishing, 1989.

Specter, Michael, "European Revolt Against Test-Tube-Altered Future," *The Globe and Mail*, July 21, 1998.

"The Spectre of a Human Clone," *The Independent*, February 26, 1997.

Steinman, David. *Diet For A Poisoned Planet.* Toronto, ON: Random House of Canada Limited, 1992.

Stwertka, Eve and Albert Stwertka. *Genetic Engineering.* New York, NY: Franklin Watts, 1982.

Taffel, Alexander. *Physics: Its Methods and Meanings.* Newton, MA: Allyn and Bacon, Inc., 1981.

Taylor, Paul, "Trans-Fatty Acids are Stealthy Health Hazard," *The Globe and Mail*, May 5, 1998.

Thompson, William. *Black's Medical Dictionary.* 31st Edition. New York, NY: Charles Scribner's Sons, 1975.

Turner, James S. *The Chemical Feast.* New York, NY: Grossman Publishers, 1970.

Warren, Susan. "DOW to Strike Key Biotech Alliances," *The Wall Street Journal* (reprinted in *The Globe and Mail*), September 8, 1998.

Whitman, R.L. and E.E. Zinck. *Chemistry Today*. Scarborough, ON: Prentice-Hall of Canada Ltd., 1976.

Williams, Roger. *Nutrition Against Disease*. New York, NY: Bantam Books Inc., 1973.

Winter, Ruth. *A Consumer's Dictionary of Food Additives*. New York, NY: Crown Publishers, Inc., 1994.

———. *Poisons in Your Food*. New York, NY: Crown Publishers, Inc., 1969.

World Health Organization. "Forty-first Report of the Joint FAO/WHO Expert Committee on Food Additives, Evaluation of Certain Food Additives and Contaminants." WHO Technical Report Series No. 837, Geneva, Switzerland, 1993.

———. "Forty-fourth Report of the Joint FAO/WHO Expert Committee on Food Additives, Evaluation of Certain Food Additives and Contaminants." WHO Technical Report Series No. 859, Geneva, Switzerland, 1995.

———. "Nineteenth Report of the Joint FAO/WHO Expert Committee on Food Additives, Evaluation of Certain Food Additives." WHO Technical Report Series No. 576. Geneva, Switzerland, 1975.

———. "Thirty-ninth Report of the Joint FAO/WHO Expert Committee on Food Additives, Evaluation of Certain Food Additives and Naturally Occurring Toxicants." WHO Technical Report Series No. 828. Geneva, Switzerland, 1992.

———. "Thirty-seventh Report of the Joint FAO/WHO Expert Committee on Food Additives, Evaluation of Certain Food Additives and Contaminants." WHO Technical Report Series No. 806. Geneva, Switzerland, 1991.

Wright, Sylvia, "Food and Your Health," *The Whig-Standard*, Kingston, ON, 1978.

Once again, the foolishness of our rulers
hath made the goodness of God of none effect.

Author Unknown

Other Great Books from Canada's Leading Publisher of Natural Health and Nutrition Books

Encyclopedia of Natural Healing

A Practical Self-Help Guide
alive Research Group
Siegfried Gursche, MH
Medical Editor – Zoltan Rona, MD, Msc

This award-winning encyclopedia is the most comprehensive self-help guide to natural health ever published in North America. It contains everything you need to heal yourself and your family – *naturally.*

1,472 pages, hardcover
Over 1,000 full-colour photos
Over 200 illustrations and charts
$69.95 Cdn
$59.95 US
ISBN 0-920470-75-0

The Complete Athlete

Integrating Fitness, Nutrition and Natural Health
John Winterdyk, PhD
and Karen Jensen, ND

Team up with a competitive athlete and a naturopathic doctor and attain optimum health by integrating fitness, nutrition and natural healing.

320 pages, softcover
Over 50 illustrations and charts
$22.95 Cdn
$19.95 US
ISBN 0-920470-05-X

Allergies: Disease in Disguise

Dr Carolee Bateson-Koch

This is the first book to explain how to achieve complete and permanent recovery from your allergies.

224 pages, softcover
$17.95 Cdn
$15.95 US
ISBN 0-920470-42-4

The Raw Gourmet

Nomi Shannon

Taste and feel the benefits of eating raw foods with *The Raw Gourmet*'s quick, enzyme-rich, high-fiber vegetable dishes, dressings and desserts.

224 pages, softcover
Over 70 full-colour photos
$29.95 Cdn $24.95 US
ISBN 0-920470-48-3

The Vegetarian Gourmet

Dagmar von Cramm

A sumptuously illustrated book featuring the best of European vegetarian cuisine.

246 pages, softcover
Over 200 full-colour photographs
$29.95 Cdn $24.95 US
ISBN 0-920470-80-7

Living With Green Power

A Gourmet Collection of Living Food Recipes

Elysa Markowitz

Reap the rewards of the living foods diet, such as proper digestion, a strong immune system and an abundance of energy.

176 pages, hardcover
96 full-colour photos
$29.95 Cdn $24.95 US
ISBN 0-920470-11-4

Fats that Heal, Fats that Kill

Udo Erasmus

Over 10,000 copies in print!

This book brings you the most current research on the common and less well-known oils and their therapeutic potential. *"A welcome relief from the one-sided approach of our current health authorities, who espouse the low-fat diet as the panacea for all health problems. The informed and balanced presentation will help to take the fear out of fats."* – Richard A. Kuin, MD, President of the International Society for Orthomolecular Medicine

456 pages, softcover
$27.95 Cdn $22.95 US
ISBN 0-920470-38-6

The Natural Physician

Your Health Guide for Common Ailments

Mark Stengler, ND

A straight-forward and practical guide for those who prefer to heal themselves with naturopathic medicine, herbs, vitamins, minerals and other natural remedies.

218 pages, softcover
$15.95 Cdn $12.95 US
ISBN 0-920470-46-7

The Cultured Cabbage

How to make sauerkraut–a practical guide

Klaus Kaufmann and Annelies Schöneck

Lovers of sauerkraut and pickled vegetables will welcome this guide to the remarkable heath-enhancing properties of lactic acid-fermented foods.

80 pages, softcover
8 full-colour photographs
$11.95 Cdn $10.95 US
ISBN 0-920470-66-1

Kefir Rediscovered!

Klaus Kaufmann

The nutritional benefits of this ancient healing food are explored. Includes recipes for kefir skin-care products and kefir foods.

96 pages, softcover
$11.95 Cdn $10.95 US
ISBN 0-920470-65-3

Silica

The Amazing Gel

Klaus Kaufmann

An in-depth exploration of the beneficial effects of silica gel on cancer, diabetes, gastritis, ulcers, skin disorders and teeth.

208 pages, softcover
$12.95 Cdn $9.95 US
ISBN 0-920470-30-0

Kombucha Rediscovered!

Klaus Kaufmann

Everything you ever wanted to know about the amazing healing properties of the fermented kombucha mushroom.

96 pages, softcover
6 full-colour photographs
$11.95 Cdn $10.95 US
ISBN 0-920470-64-5

For the Love of Food

The Complete Natural Foods Cookbook

Jeanne Marie Martin

In this updated version of Martin's classic book, you'll find comprehensive guidelines for food combining, vegetarianism, fasting, vitamins and minerals, bread making, sprouting and cooking with herbs.

384 pages, hardcover
13 full-colour photographs
$29.95 Cdn $24.95 US
ISBN 0-920470-71-8

Healing with Herbal Juices

Siegfried Gursche MH

The first book of its kind, *Healing with Herbal Juices* presents a simple, effective way to benefit from the superior healing power of herbs.

156 pages, softcover
$18.95 Cdn $16.95 US
ISBN 0-920470-34-3

Return to the Joy of Health

Natural Medicine & Alternative Treatment for All Your Health Complaints

Zoltan P. Rona MD, MSc

The most up-to-date, comprehensive guide to natural healing by a medical doctor.

408 pages, softcover
$24.95 Cdn $19.95 US
ISBN 0-920470-62-9

A Diet for All Reasons

Paulette Eisen

More than a cookbook, *A Diet for All Reasons* explains why a meat-, egg- and dairy-free diet is essential for cardiovascular health, reduced stress levels, and overall well-being.

176 pages, spiral bound
$15.95 Cdn $12.95 US
ISBN 0-920470-68-8

Devil's Claw Root and Other Natural Remedies for Arthritis

Rachel Carston
(updated and revised by Klaus Kaufmann)

The symptoms of arthritis can often be completely eliminated by taking devil's claw root. Find out how in this down-to earth guide.

112 pages, softcover
$11.95 Cdn $9.95 US
ISBN 0-920470-36-X

Menopause

Time for a Change

Merri Lu Park

An in-depth and practical guide that will put you in control of your life and your health.

304 pages, softcover
$18.95 Cdn $16.95 US
ISBN 0-920470-33-5

The Breuss Cancer Cure

Rudolph Breuss

The renowned Dr Breuss offers practical advice for the prevention and natural treatment of cancer, leukemia and other seemingly incurable diseases.

112 pages, softcover
$13.95 Cdn $11.95 US
ISBN 0-920470-56-4

The All-in-One Guide to Herbs, Vitamins & Minerals

Victoria Hogan

The All-in-One Guide offers the most up-to-date information about the many healing herbs, vitamins and minerals available today.

112 pages, softcover
$7.95 Cdn $5.95 US
ISBN 0-920470-99-8

alive books are available at health food stores and book stores or by calling (800) 661-0303 or (800) 661-6513.

To The House Of Commons: Petition for Mandatory Labelling and Thorough Testing Of All Genetically Engineered Foods

We, the undersigned, being over the age of 18 and citizens of Canada, call upon Parliament to legislate clear labelling on all genetically engineered seeds, foods, and their by-products available in Canada and further that these products be banned from the market until they have undergone rigorous long-term testing to prove their safety when consumed by humans, and when they come into contact with all other species with whom we share the planet.

Signature	Print Name	Address

Please return completed petition to:
Consumer Right to Know Campaign
500 Wilbrod Street, Ottawa, ON K1N 6N2
Tel: 613–565–8517, Fax: 613–565–1596
Email: rwolfson@concentric.net
http://www.concentric.net/~rwolfson/home.html